To: Cho
Love, Roger

A Dangerous Return:

Surprising Lessons from the Congo

A Dangerous Return:

Surprising Lessons from the Congo

By Roger L. Green

Xulon Press

Xulon Press
2301 Lucien Way #415
Maitland, FL 32751
407.339.4217
www.xulonpress.com

Printed in the United States of America.

ISBN: 9781545603703

Dedicated to the memory of my sweet wife,
CAROL S. GREEN,
without whom this book would not have been possible.

TABLE OF CONTENTS

Acknowledgements Xi

PART ONE: THE EARLY YEARS IN KIVU, BELGIAN CONGO 1950-1960

Chapter One: Death In The Window! 3
Chapter Two: Adventure Begins 5
Chapter Three: Africa! 9
Chapter Four: Monkeys, Africans, And Houses 14
Chapter Five: Diseases, Snakes, Scorpions, And Attack Ants! 19
Chapter Six: Becoming "Congolese" 26
Chapter Seven: Jansen Hall 29
Chapter Eight: Crazy Stuff! 43
Chapter Nine: Crazy!—And Not So Crazy 49
Chapter Ten: Learning The Language. 57
Chapter Eleven: Traveling To America. 61
Chapter Twelve: Congo—The Eye Of The Storm!. 64
Chapter Thirteen: Danger Ahead!. 69

Chapter Fourteen: The Winds Of Change 72
Chapter Fifteen: Trouble At Our Doorstep(S)! 76
Chapter Sixteen: Fleeing Again 80
Part One: Kivu Years Images 90

Part Two Transitions In America Images 1960–2011

Chapter Seventeen: Safe At Last. 99
Chapter Eighteen: Life In The United States 102
Part Two: Transitions in America Images 1960–2011 104

Part Three: New Life Goals And Uganda Missions 2011–2012

Chapter Nineteen: Alone After 45 Years. 109
Part Three Uganda Missions Images 2011–2012 113

Part Four: Unlikely Return To Congo DRC 2012

Chapter Twenty: A New Vision. 119
Chapter Twenty-One: Africa, My New Goal!. 121
Chapter Twenty-Two: Conflict And Contrast 122
Chapter Twenty-Three: Arriving "Home" 126
Chapter Twenty-Four: Reunion In Congo 130
Chapter Twenty-Five: Catching Up 135
Chapter Twenty-Six: Nostalgia In Bukavu 138
Chapter Twenty-Seven: The Big One 145

Chapter Twenty-Eight: Extortions And Bribes 147
Chapter Twenty-Nine: Night, Dangers, And Darkness 152
Chapter Thirty: Rain, Rain, Rain! 159
Chapter Thirty-One: Hassles And Obstacles 163
Chapter Thirty-Two: Church In The Jungle 165
Chapter Thirty-Three: Miracles Do Happen! 168
Chapter Thirty-Four: Congo Strikes Again 174
Chapter Thirty-Five: Arrived 178
Chapter Thirty-Six: Leaving Downcountry Congo DRC 189
Chapter Thirty-Seven: Flight From Hell 193
Chapter Thirty-Eight: Most "Hellacious Road" 197
Chapter Thirty-Nine: A "Toilet Tsunami" 203
Chapter Forty: Amsterdam To Detroit .206
Chapter Forty-One: Home At Last! 208
Part Four: Unlikely Return To Congo DRC Images 2012 . . . 209

RETROSPECTIVE: A DANGEROUS RETURN: SURPRISING LESSONS FROM THE CONGO

ACKNOWLEDGEMENTS

MANY THANKS TO my late wife, Carol, our late daughter, Nicole, and my children and their spouses, Michelle, Danielle, and Tony, and especially to my wife, Linda. Simply put, without Linda's support, love, and encouragement, this book would not have happened.

A very special thank you to Buyamba, Inc., of California for making it possible for me to go on mission with them to God Cares Uganda. Special thanks to donors who supported my mission.

Also a special thank you to my church, Cross Timbers Community Church, Argyle, Texas, and in particular to Dave Schmille for making it possible for me to extend a Uganda mission trip to return to Congo DRC, where I spent ten formative years of my life.

Tom and Kathy Lindquist were my protocol, hosts, confidants, and leaders for getting me into Congo DRC. Thank you so much!

Jim and Louise Lindquist gave invaluable input to this project. Thank you!

Mrs. Marian Ammons and Mrs. Jeannette Rudder were invaluable with advice. Mrs. Linda Coolen and Mrs. Marjorie Wall were also great helpers. Thank you!

Finally, my brother David and his spouse were wonderful in providing documents, photos, stories, memories, love, and support. Many, many thanks!

Thanks to my proofreaders: my wife, Linda Spaugh Green, and my daughter, Danielle Green Lackey. Thank you to my editor, Charlene Patterson.

Thank you to Albert and Reba Jansen for their permission to use their book, *The First Six: Early Years of Berean Mission in the Congo*, as a source.

Thank you to my late mother, Hallie L. Green, for permission to use her book, *No Certain Dwelling Place: Forty Homes in Forty Years*, as a source.

Thank you to my late father, Ernest L. Green, for permission to use his book, *Uncle Ernie's African Stories*, as a source.

PART ONE

The Early Years in Kivu, Belgian Congo

1950-1960

CHAPTER ONE

Death in the Window!

AS A TENDER eight-year-old boy I walked up to an open window in our house, looking out to see my friends playing outside. As I did so, I laid my hand on the window ledge to steady myself. In the blink of an eye, I felt something strangely cold, rough, and slimy under my hand. *Wow!* I realized as I jumped back, a snake was coiled there, sunning itself and sleeping. Running to my father's study, I yelled, "Snake, snake—help, help!"

Dad answered, "Oh, Roger, there is not a snake."

I retorted with trembling hands. "Yes, there *is*. Come and see for yourself."

Dad came with me back to the window and saw the huge snake, a relative to the black mamba, lying on the window ledge. It had been awakened by my touch and was poised to strike. Looking at the snake, we realized we were dealing with a dreaded mamba. The Africans call it "Two Step," since after it bites you, you have two steps until you die.

Hearing the commotion, an African rushed to the window, bravely yelling, "Where is the snake?" Seeing the snake, he immediately panicked and ran out of the house, leaving us with the snake.

A second African came bounding into the room, full of bravado, and then he let out a yell and disappeared. We still had "Two Step" to deal with.

The snake was watching us intently. If we tried to move in, he would strike. If we tried to hit him from a distance and missed, he would escape into our house.

My brother David had arrived with a big, long stick. Frightened, Dad said, "David, give me the stick. I will try to kill it myself. Stand back, I don't know what will happen." Raising the stick over his head, Dad took aim.

Just then a third African rushed in, immediately taking charge and saying, "Give me the stick!" Grabbing it, he took an angled step forward, bringing the stick down from above his head, crashing it onto the curled-up body of the snake. As he held down the snake, one of us grabbed a machete, ran to the window, and cut off the snake's head. With a shout of joy and victory, we pulled the snake from the window ledge and stretched it out onto the floor. It measured over six feet long. Thank God it was dead.

CHAPTER TWO
ADVENTURE BEGINS

OUR ADVENTURE BEGAN in 1950, when my family of six moved most of what we had to a nation called the Belgian Congo, in the heart of darkest Africa. What began in 1950 ended in 1960 with a hail of bullets and crippling fear. As the year 1950 began, my parents, Ernest and Hallie Green, were answering God's call upon their lives to go to Africa, under the Berean Mission, St. Louis. Dad was an evangelist/Bible teacher and a translator. Mother was a registered nurse. My oldest brother, Eldon, age nine; my second brother David, age seven; myself, not quite five years old; and our baby sister, Patty—who was in Mother's arms—were all set to go. The Belgian Congo was owned by the country of Belgium, as a colony of Belgium. The colonization of the continent of Africa was in full swing during those times. The Congo is five times the size of Texas, with over two hundred African tribal groups and languages.

The Belgian government ruled over the Congo. They built roads, developed industries of mining, rubber, and logging, and created a system of schools and regional medical facilities. While the regional centers were fairly well developed, they were very primitive, and in some cases barely useable. Malaria, typhoid, pneumonia, yellow fever, jaundice, hepatitis, and a thousand other tropical diseases ran rampant, not to mention snakes, scorpions, gorillas, baboons, leopards, and other dangerous creatures! Witchcraft, witchdoctors, curses, tribal conflicts, polygamy, and even killings were commonplace. Our safety was clearly in the hands of the God who had called our family to Congo, Africa. The missionaries' task was to bring the Gospel of God's love to the native villagers of Equatorial Belgian Congo.

After an arduous trip from Nebraska to New York City, we excitedly boarded a German ship, a freighter named the SS *Steinstradt*. We were about to experience adventures like we'd never imagined before. As we pulled away from the dock in New York Harbor, we slid right past the Statue of Liberty. Dad stood on the outside deck, crying for hours from the thought of leaving family, friends, and country behind, until finally the lights of the United States were so far behind us that there were no lights at all. I vividly remember standing close to him, trying to give him some kind of comfort, even though I was only four years old.

From Mother's book, *No Certain Dwelling Place: Forty Homes in Forty Years*:

"Our seventh day at sea was very windy and stormy. Going to the dining room on board ship, we saw the tables had sideboards around all four sides. It was a good thing, for as the ship would roll from side to side, our plates would slide rapidly away from us. Dad was quite seasick and couldn't come to the dining hall at all!"

Finally, the storms abated. With endless days and nights at sea with no land in sight, we boys lost several softballs, baseballs, and gloves "accidently" tossed overboard.

Again from Mother's book:

"Sailing South, it grew warmer and warmer, so to break the monotony guided tours of the *S.S. Steinstradt* were given.

One time we went down to the sailor's quarters and then below to the huge engine rooms. Another day we went up to the bridge atop the ship, where the boys were allowed to steer the giant ship.

Hearing shouts of 'Come and see' we ran to the deck and saw—it was the Coast of Africa, the lower part of the 'hump.' We would soon turn East toward the coast at a point where the giant Congo river empties into the ocean. The ocean water was taking on a different hue. No longer was it alternately blue then green, it was now a dirty brown color. We were seeing the effects of the mighty Congo river, over two hundred miles out to sea. At 6:30 a.m. April 12th, we arrived at the mouth of the river. A small boat came alongside containing the man who would pilot our great ship up the river to the port of Matadi. Slowly going up the river we saw a man walking along a path, carrying an umbrella to protect himself from the hot sun. Behind him came his wife with a heavy load on her back and nothing to protect her from the sun. Following her were two children, completely naked." *This is Africa!*

Our family, fresh from Nebraska and the Heartland of America, was beside ourselves with excitement. We were ready to dive into the middle of this new adventure. After a twenty-one day voyage from New York City, our family could not have been any "greener," raw rookies, in fact, certainly living up to our family name. In fact, since Mother's name was Hallie, the three of us brothers had earned the nickname "Hallie's Comet"!

The ship docked at Matadi, Africa, and Dad and his boys were allowed to leave the ship to walk around. Patty and Mother stayed onboard a bit and were then told it was too late for them to disembark. Many Africans walked by, and baby Patty smiled at all of them. This made the Africans very excited to see her smile. Finally, after twenty-one days at sea we were in Africa! We were in hot, sticky, smelly, strange, and wild Belgian Congo, Africa.

CHAPTER THREE

AFRICA!

AFTER WE DISEMBARKED on the coast, there came days of driving on barely passable roads to the capital city of Leopoldville (today called Kinshasa). The tropical sun, heat, and humidity shocked us immediately. Among our earthly possessions was a 1950 Ford, donated by friends in America. I vividly remember watching the huge crane unload our car from the bowels of the ship! Our traveling companions onboard ship were a British doctor and five venomous snake hunters from Florida. We twelve passengers on that freighter made quite a diverse group. After catching our breath for a couple of nights at a missionary guest house in Leopoldville, we booked passage on a Sabena Airlines DC-3, to the interior regional border post of Bukavu, Congo. How could I know that sixty-two years later I would again stand on this very spot?

Upon arrival in Bukavu, we were met by veteran missionaries waiting to drive us on our trip downcountry over the infamous "Route de Kimbili," climbing mountains, skirting cliffs, and traveling an average of perhaps five to ten miles per hour. In those days the "road" was really a sand or clay track hewn out of the tropical rainforest. On good days it was hot, dusty, and full of pot-holes. On rainy days it was wet, slippery, and splashing with reddish-brown, muddy gook. There was scarcely room for one car, let alone room to meet and pass oncoming vehicles. Every day, we would encounter Africans walking with heavy loads on their heads or strapped to their backs.

Fortunately, we might see only one or two vehicles in a full day's drive. When we did, the vehicle traveling uphill had the right of way. The vehicle heading downhill had to find a place wide enough for both vehicles to pass, or back uphill until such a place was found. For a good distance, the "road"—an overgrown path—was one way only, so every other day travelers could proceed in that one direction. If one wished to take their lives further into their own hands, on Sundays travel was allowed in both directions. Road hazards included sheer drop-offs, landslides, trees fallen across the road, violent thunderstorms, bridges washed out, and deep holes to avoid. We dodged animals running across the road. If the road was blocked by a landslide or a large tree, we would honk the horn to arouse some local villagers to help us cut or dig our way through. For a gift they would give us a helping hand.

Our gift to them was usually something like fifty cents per man, for hours of backbreaking work. In the 1950s that was a huge wage for the Africans, who were excited to get it. We drove the next three arduous, sometimes terrifying days to our first posting, at the Berean mission station in Katanti, Belgian Congo, Africa, set up on a hill in the middle of the jungle. Sixty years later, one major house was still there. The mud-and-stick church building is still there today—and holding weekly services, but I am getting ahead of the story.

Dad's note, from Albert Jansen's book, *The First Six: Early Years of Berean Mission in the Congo*:

"Upon traveling through miles of mountains, we entered the great Kivu forest. A very joyous welcome awaited us at our first mission stop, Ikozi (meaning "the rock"). We stood on the steps of a stone house as the Africans danced and sang us the welcome song, *'Twa Bogaboga.'* Our hearts and eyes overflowed with joy listening to them sing another song, "There Is A Fountain Filled With Blood." We knew the tunes even if we did not understand the language, a tribal dialect called *Kilega*. Right after lunch, we headed down the road to our own mission station, *Katanti*. The drive was only about eighty miles, and much easier than what we had been through. What a welcome awaited us at *Katanti*!

As for me, I can vividly remember that as we neared the mission, hundreds of Africans came out of their huts to stand along the roadway, smiling, cheering, and calling out the missionaries' names. Many men were in ragged shorts, with no shirts. Women were in a rudimentary skirt tied at the waist with a string—topless, as was their custom. Children were often naked.

Climbing the hill we were met with a jubilant celebration given for us by the Africans and the few white missionaries already working at that post. What a moment for us newbies! I well remember hiding behind my mother's skirts.

Dad's recollections of that moment, from Albert's book:

"We had a supper of rice, African green vegetables, mystery meat, and water. The Jansens, (who founded this post in the late 1930's), were our hosts. We thought we were eating wild pork but Mamie Jansen informed us it was aardvark! How happy we were to walk into our own house prepared for us at *Katanti*!"

Our family could not help but reflect upon the amazing and difficult issues the original six Berean missionaries had encountered upon their entry into the Congo in 1938. Those six brave Americans had left their families and most of their earthly possessions behind, to answer God's call to take the Gospel to the mostly undeveloped jungles of Belgian Congo, Africa. In contrast to our first trip inland, their trip was trekking in on foot, in canoes, and in rickety river boats, with no roads nor transportation in many areas. The area they were called to serve was undeveloped except for a few Belgian mines, and many of the Africans had never seen a white man. In addition to a language barrier, cultural differences, and strange-looking food, those six slept in tents as they hacked their way through the jungle filled with dangerous animals. Those trails had to be cleared by hand, as did the property for the very first mission post, on top of a mountain called Musuku.

Those strong, committed people braved malaria, amoebic dysentery, poisonous snakes, leopards, deadly sleeping sickness, and other unknown tropical parasites. That was just the beginning of their woes. In the 1940s, polio swept through that region of the Congo, and the mission post was quarantined and on lock down for quite a time. Although travel by ship to and from the United States to Europe and Africa was extremely dangerous, other early missionaries braved German U-boat-infested Atlantic waters to travel to their missions in Congo. Significantly, none of their boats were attacked.

A few years later, the Belgians began to open a road into the jungle area for the first time. Wanting to be near the new road, the mission at Musuku was moved to the Katanti mountaintop. Building and clearing began all over again. Moreover, several times the new mission post was shut down and the missionaries had to flee the area. Unrest among the Africans, a lack of honest workmen, and even death threats to the missionaries from a neighboring tribe made the post very dangerous, until order was restored. Lastly, during World War II and its aftermath, supplies were on a ration basis, including food products, gasoline, and other perishables. Into that setting my family began its work with the African tribe called the Balega, in Belgian Congo in 1951.

So began ten years of mission work, living in the undeveloped provinces of Maniema and South Kivu, Belgian Congo. Although I wouldn't trade the experience for anything, the ending was pretty "sketchy," to say the least. But first, we had to adjust and learn the ways of the Balega tribe and Africa. My siblings and I were growing up "Congolese." Here's how those adventures went down.

CHAPTER FOUR
MONKEYS, AFRICANS, AND HOUSES

GROWING UP CONGOLESE looked like this: Upon our arrival, we moved into a large house made of poles stuck into the ground, then lashed together with a native vine, then a mud plaster to make walls. The plaster was then coated with whitewash. Inside there were no ceilings, just open to the rafters. The roof was wooden poles used as rafters, with bales of palm leaves lashed to the rafters, to form a thatch roof. The house did have window openings, but no windowpanes or coverings. For those, we used either hanging cloth or semi-clear plastic sheets. No closing windows per se, just a covering to keep out the rain during the rainy season. For doors, we had just one homemade wooden screen door. For beds, mattresses were placed on a wooden frame on the dirt floor, without a head or foot board. Over the top a mosquito net was hung from a rafter. The mosquito net was intended to ward off malaria, mice, rats, monkeys, lizards, scorpions, snakes, and who knows what else.

The floors were hard-packed dirt. Thus, the house was built like an African house, only much bigger than usual.

On all four sides of this particular house, the roof extended out about eight feet to form a covering over a veranda. This gave the house almost a plantation-style look. The veranda roof was supported by poles at equal intervals, and around the veranda itself was a ditch to catch the runoff of the rain. There was no running water, nor gas for cooking. The bathroom had an old toilet, with a drain dug under the walls to a pit outside. Water had to be carried daily to barrels on a platform just outside the house, to provide gravity feed to the toilet, kitchen, and rudimentary sink.

For refrigeration, someone had found a very primitive unit called an "IcyBall." It was a large insulated cabinet with a divider in the center and two lids that would open upward. On the right side of the container was a large steel ball attached to a crossover to another steel ball on the left side. The ball on the right had a fire pit under it in the container, for a small kerosene-fueled fire. The fire was placed to heat the right ball. Through some chemistry I never understood, the heat would be exchanged to cold on the ball on the left side of the container. The smaller left-side ball would radiate cold, and leave space to store things below—things we needed refrigerated. Primitive, yes, but quite functional. I have never seen another "IcyBall."

For cooking, we had an old wood-burning stove, with a vent just up through the rafters and the thatched roof. Somewhat dangerous and smoky, but we thought, *This is Africa!*

Since we had one of the largest families among the missionaries, our house was one of the two biggest on the missionaries' compound. There was another similar house holding a large missionary family, and a third smaller house for a missionary couple. There was also an older couple from the original six American missionaries who had come first to open up a mission in the jungle in 1938.

In the 1950s in Congo, the custom was to hire African "houseboys" to help the family with cooking, washing, laundry, and so forth. Shortly after our arrival, we hired Wamenya Avoca, a cook, and another man whose name I've forgotten. He did laundry and other cleaning jobs. Wamenya Avoca became a real favorite of ours.

The rest of the mission compound consisted of a carpentry shop, an African school building, a church, and a fenced-in area we dubbed "The Fence." The fence was for young African girls who came from the bush villages to the African mission school, to have a safe compound and living space away from the general population. This kept them safe from outsiders and villagers alike, including curious young missionary boys. Near to the fence was a large mud-and-stick building that housed the mission church. Halfway down the hill was the African village, complete with a large soccer field. The African village had elders, preachers, women and kids, workmen, and others. Across from the African village was a mud-and-stick dispensary building where my mother, a registered nurse, held a clinic every day—and also 24/7. As I recall, she had two African helpers she trained to help her.

Below the dispensary, the road ran down the hill from the mission compound to the main "road" below. For transport we had brought our old 1950 Ford from America, and each of the other two missionary families had donated American vehicles, as well. One of the vehicles was a pickup truck with the name "Shasta." We always said, "She hasta have gas, she hasta have oil, and she hasta have tires!" Gasoline, oil, tires, and kerosene were available only a long drive away, at the regional center, Shabunda. If they weren't available there, we had to make a longer trek to a larger town called Kindu. Therefore, we stockpiled fuels in steel fifty-five gallon drums.

The missionaries all became, for better or for worse, mechanics of a sort. Many, many times parts were made from pieces of metal, pins, needles, even the garters from missionaries' girdles! Since nearly everything had to be imported from across the sea, and then by land to our post, wait times for parts, accessories, and supplies could reach three to six months. In those days, freight transported by air had not become a workable option. Sea, land, or a runner by bicycle or on foot were the only ways to transport things to us, in country.

One of our first challenges was to get used to African food. Our diet included manioc, plantain, rice, greens (*sombe*)—made from a special type of grass—and *bugadi*. Bugadi was a gray mealy spongy concoction, boiled and then cooled into a big blob. It never had much nutritional value, nor taste—rather, it was a staple used for a filler. (Bugadi is still used to this day, under various names in different African countries.) For meat, we ate antelope, bush wart-hogs, pigs, chickens, monkeys, goats, canned corned beef from Australia, and certain birds. We also raised rabbits for meat and grew meager gardens for vegetables. Papayas, tangerines, avocadoes, palm nuts, pineapples, and peanuts were readily available fresh. The meat was most often killed and provided by the Africans. Because it was often trapped and left a while to die—or sometimes killed with poisoned arrows—Mother had to be sure the meat was bled out properly, and any meat affected by poisonous arrows was cut out and discarded. As for the Africans, they ate almost anything at any time or place, regardless of contamination. I often thought their stomachs were made of iron. Sometimes they liked the grossest concoctions, including the entrails, the best! Once a year, the missionaries would pool their meager resources to buy and slaughter a very tough and scrawny cow from an African driving a small herd of five or six through the jungle forest.

For breakfast we usually had a water-based powdered oatmeal, banana, papaya, or grapefruit. We used powdered milk called "KLIM," which came from Belgium. Not very tasty, but it filled the need. My mother taught our cook to bake bread, so we sometimes had toast browned over an open fire. Dishes and silverware had come with us from the United States.

Water had to be boiled for thirty minutes before we could drink any of it. This was to ward off dysentery, E. coli, and a million other stomach and digestive problems. Our African cooks had to be watched to make sure they were boiling our water. They could not understand this need, since they could drink any water, any time, anywhere. Actually, we boys took to the African ways and pretty much drank any water, any time, too. Undoubtedly that is why the native Africans named me "*Banuamazi*." My name translates literally into English like this: "He drinks water from a stream or river, without knowing who or what pooped or peed upstream!" Without question this was why we got worms and malaria so often, as well. Boys will be boys.

CHAPTER FIVE

DISEASES, SNAKES, SCORPIONS, AND ATTACK ANTS!

SPEAKING OF ILLNESSES, Africa offers quite a smorgasbord. Before leaving the States, we had all been inoculated against yellow fever, hepatitis, jaundice, malaria, meningitis, polio, and other exotic diseases. However, upon our arrival in the jungle, we found we needed to take quinine as an anti-malarial. The dosage was either one pill a day, or more often, one per week. Did this guarantee no malaria? Not at all! We got malaria about as often as one contracts a cold in the U.S. What the quinine did do was keep the malaria from reaching dangerous levels of infection and fever, which could easily be lethal. Somehow, we made certain to never run out of quinine over our ten years in the Congo.

In addition, there was something called filaria. This was a small, parasitic worm-like creature, which would grow in creeks, rivers, and ponds. They entered one's body through soft tissue and then moved around under the skin. As they moved, they caused a huge itching sensation. The only medication caused strong aches and pains in one's joints, or the Africans will take a fishhook and pluck the worm out of one's eyeball as it crosses the front of the eye. Dangerous, yes! Effective, yes! When that little filaria worm crossed one's eye, it produced such itching, one would try almost anything to get rid of it, believe me.

Pink eye was problematic too, as many of the Africans suffered from it, as well. And it is highly contagious. Stomach worms came from eating foods not thoroughly washed by our houseboys and cooks. All of our family had to take castor oil once every six months, for de-worming. Ugh, what a group event that was.

For the numerous scrapes, cuts, and abrasions we received, Mother always kept plenty of iodine or Mercurochrome on hand. There's nothing quite like pouring pure iodine into an open cut. *Yow.* Due to the heat and humidity, athlete's foot was very common, as well. For that, we used Absorbine Jr. from America, which will definitely light a fire.

In spite of a "million" inoculations we had received before we left the U.S., Mother suffered a terrible bout with hepatitis while we were on a trip away from our Katanti station. She was diagnosed with a "difficult case," and her life was definitely in danger. In such a precarious condition, she had to stay at a far-off mission station, Kamulila, to be nursed back to health by a second missionary nurse, Betty Lindquist. Betty was the wife of one of my parents' colleagues. Many, many prayers went up for her recovery, which took around three weeks.

We boys and little Patty had to return to our school and work, a two-day drive from where our mother lay battling for her life. Our only communication with her was by shortwave ham radio, once each week.

The only things Dad knew how to tell the African cook how to make was rice and peas. Yes, we had rice and peas two meals a day for three weeks. Mother's nurse Betty Lindquist's husband, Irving, had been one of the six original Berean missionaries into the Congo in 1938, and he and his wife became great mentors for my parents. How could I know that fifty years later I would reestablish a great relationship with their son Tom, and see him again, on Skype and then back in the Congolese jungle?

Early on we learned all about poisonous spiders, poisonous scorpions, and poisonous snakes, everything from the black mamba, to the cobra, to pit vipers, horned vipers, and more. Since our houses were made with mud, sticks, and leaves, they made a perfect habitat for all the things that might hurt us. Snakes, mice, and lizards shared our homes with us. We got some cats from somewhere to help with the rodent population.

One of my favorite scorpion stories comes from my father's book, *Uncle Ernie's African Stories:*

"It was bedtime. Everyone was taking turns using the bathroom. It was Lydia's turn to use the facility. (Lydia was a rookie missionary nurse who had just arrived from America to join us, and was staying in our house.) Sitting on the toilet she reached for the toilet paper when wham—she felt a sharp sting on her finger. Looking quickly she saw the telltale straight body of a scorpion, with its curved-up tail stinger. Pain rushed through her finger, into her hand, and up her arm. She called to Mrs. Green, 'I've been stung by a poisonous scorpion!' Quickly cutting an *X* on the spot of the bite with a razor blade, Mrs. Green tried to squeeze the blood from the area in an effort to draw off the poison. Even with this fast response, Lydia suffered with lots of pain up her arm for some time. After a time, we were relieved that her body had defeated the sting!"

Black mamba snakes were frequent visitors inside and outside our home. The mamba is very dangerous. Its venom will kill you in a matter of minutes. We learned to live in a constant state of awareness and readiness. I consider it no less than a miracle that in ten years in the Congo, I never knew one missionary who was bitten by a black mamba. Sadly, I can recall Africans who were bitten, and some who lost their lives. Sometimes they died from waiting too long for treatment. Often they first went to their native witch-doctors with their strange concoctions of herbs, chicken blood, mud packs, chanting, and calling on devils. Sometimes they could have been saved if they'd first come to Mother's dispensary for the proper treatment. Old habits are hard to break.

Again from Father's book, *Uncle Ernie's African Stories*:

"'Snake! Snake!' Africans came running out of the small mud-and-stick thatched washhouse behind our house, darting in every direction. Jumping up from my desk, I ran toward the washhouse, where men had been putting on a new thatched leaf roof.

'Where, where?' I yelled.

'There! There!' they shouted, pointing toward a large, round, curled black object on an exposed rafter, a large black mamba. Just then, Ngandu, the African evangelist, came running up. Ngandu sent someone running to his house for a long sharp spear. Meantime, everyone stood still, watching the mamba. When the man returned with the spear, Ngandu took aim, and zinged the spear through the air. It slammed right into the heart of the large black spot and stuck into the rafter. As the Africans shouted, the long black mamba slowly uncurled with half of its body outside the rafter and the other half inside. The half with the head was on the outside, eyes flashing, and tongue darting in and out of its mouth like lightning.

Be careful!' shouted Ngandu. 'It can still bite you and kill you!'

As the snake thrashed about, an African grabbed a long, sharp machete, ran up to the snake, and pinned its head against the rafter. Then he cut off its head. The danger was over; the mamba was dead. I marveled at how Ngandu could throw his spear nine feet with the accuracy to hit the mamba. The scare was over."

Author's note: A poisonous mamba can bite down and emit deadly venom even with its head cut off. One must still be careful. As for me, I had no idea that I would again meet up and remember this story with Ngandu almost a lifetime later. Ngandu was a wonderful friend.

Then there were the dreaded "attack ants." Sometimes they're called "soldier ants." I do not know the scientific name for these large red ants. What I do know is how they messed up our lives many, many times.

From Dad's book, *Uncle Ernie's African Stories*:

"Attack ants live in deep holes in the forest; a hole running over with large ants. When they march, they go four or five abreast through the jungle, the road, or a clearing. There are 'sergeants or captains' along the side of the moving column every few inches to keep them in line. Other ants form a cover over the moving line, forming a kind of tube with a roof of ant bodies. Through the tube go millions of ants, and the queen. If someone comes along and messes up the tube-like structure, the many 'captain' ants will quickly see that it is restored. When the colony of ants arrives at a garden, or into a home where there is food, they spread out everywhere for the 'attack.' These attack ants have very sharp pinchers, with which they pick off flesh—any flesh; the flesh of cockroaches, snails, mice, rats, chickens, animals, and even humans.

One time we were sleeping in the owner's home at another mission station called Ikozi. Many missionaries and Africans were there for the Annual Conference. A visiting language specialist, (who had come to observe the proceedings), had a room next to ours. Early one morning we were awakened by a weird, sharp ticking sound. Then came an anguished cry, 'Mr. Green, I'm being bitten!' Grabbing a flashlight I [Dad] chased away the darkness. The floor to our guest's room was covered with a red carpet of attack ants. Yelling to him, 'Get out of your room,' we grabbed our robes and ran into the living room. Seconds later our guest joined us, but he kept jumping, grabbing at himself, and twisting in a frenzy. 'You shouldn't have put your clothes on,' I said. 'They are full of ants.'

Running to another room, he quickly undressed, carefully shaking the ants from his clothing and picking them off his body. Finally, he returned to the living room. There was no more sleep that night. The houseboys came and using a mixture of ashes and DDT spray, (a form of very toxic spray now banned in the United States), turned the ants around and finally out of the house. Going to breakfast someone said, 'Such is life in the Congo.'"

Many, many times in my youth, we would be invaded by attack ants.

Often, the best thing to do was to leave the house to the ants, letting them pick over whatever they wanted. Then they would leave after a few hours, until a later date, when we least expected them. Another way to rid your place of attack ants was with gasoline. Once, as a young boy, I poured gasoline all along our house on a solid line of the ants. To my dismay, the gasoline flowed downhill and right into a cooking fire used by the Africans. Sadly, the cook shack went up in flames. To be honest, the ants were pretty good at housecleaning, although a great nuisance.

CHAPTER SIX
BECOMING "CONGOLESE"

SOMETIME AFTER WE arrived and settled into our new lives, Mother was asked to start a school for missionary kids, on our mission station. Since she had a teaching degree from Nebraska, and since more and more missionary families were arriving on the Congo mission fields, she was the most qualified person to do the job. In those days, one's workload wasn't really a big consideration. Although she ran the dispensary and clinic 24/7, and taught an African ladies' Bible class, she accepted the position of the first teacher, superintendent, and principal of Berean Christian Academy missionary kids' school—on top of her other jobs.

The curriculum was based upon a highly recommended correspondence school program from the Calvert School in the U.S. We didn't do the correspondence part since Mother had all the training to carry out lessons, testing, and so forth. However, we did receive books, lesson plans, and supplies from America. Often, these would be carried to Africa by someone coming or going from furlough. Berean Christian Academy had grades K–12, all in a large, one-room school building. Since she had taught in similar schools in rural Nebraska, Mother had good experience to lean upon.

For Dad's part, he was trained in the United States as a seminarian, preacher, and Bible teacher—but was asked to supervise the building of a new mud-and-stick school building for the missionaries' kids. Having had no building experience at all, he supervised the building of the school, which came out leaning like the Leaning Tower of Pisa! Fortunately, the men were able to brace the structure so all was not lost, and we moved in for school.

Our school day was managed entirely by Mother, like this: At the front of the class was a section with eight to ten chairs, facing toward the teacher's desk. As the hours went by, each class, K–12, would have their own exclusive time to come to the front, take their seats, and do their recitation or blackboard work for the teacher. Then the teacher would teach them their lessons for the next day's work. At the end of their time, that specific class would return to their seats, and their places at the front would be taken by the next highest class. The classes not in the front with the teacher were meant to do their schoolwork for the next day. So it went throughout the days. There were morning and afternoon recesses, with a lunch break in the middle of the day. As I recall, school would dismiss around 3:30 p.m., if one had done their work satisfactorily for the day. I remember many times when I had to stay late after school, since I had spent time goofing off instead of doing my work.

Someone had found some window screens and put them over the school windows. I well remember the day a beautiful hummingbird flew right into one of the screens, getting stuck with his long, curved beak through the screen holes. Since I loved to hunt birds,

I was appointed the privilege of going outside and pulling the bird out of the screen. I then let him fly away.

At this point, I want to mention the wonderful scholastic record of that small missionary kids' school in the heart of the jungle. Any and all students who would return to the United States would always be put a grade ahead of their age level in America. Such was the great quality of the teaching in Berean Christian Academy. As word of the school got out to other parts of the Congo, quite a number of other mission organizations requested that their kids come to Katanti and study at the Christian school. Negotiations ensued, with the result that our small school became larger. And those new students needed somewhere to live during the school year, since their parents were assigned to mission posts many miles away. Thus began the story of our house becoming a dormitory for missionaries' kids. Jansen Hall, made of brick, became a home away from home for missionary kids at Berean Christian school.

CHAPTER SEVEN
JANSEN HALL

JANSEN HALL WAS a large, two-story brick house that had been started by Albert Jansen, for himself and his wife, Mamie. Albert was a trained builder and desired to build a "dream house" for his wife. For many months, the new brick house was under construction. All the bricks had to be made by hand and fired in a brick kiln, and the wood beams had to be hewn by hand, from large trees in the jungle. As cement was sent in from America, it too was mixed onsite as necessary. The dream house was taking shape. Alas, when the house was only half finished, Albert and Mamie Jansen had to return to the U.S. due to a serious illness that had beset Mamie. Sadly, word came later that Mrs. Jansen was so ill, she could not return to Africa, the place she loved. The work stopped, then was slowly resumed under the supervision of Dad and another missionary, Neil VanderPloeg.

Despite not having the expertise of Albert Jansen, Dad and Neil were able to make progress on the dream house, even though some short cuts had to be made in the construction. Finally, the brick house with the tin corrugated roof was finished, just in time for our family to move in at Christmastime. The floors were concrete on the first floor and wooden on the second floor. The second floor's flooring was never finished correctly, resulting in three to four inch cracks between each floorboard. We lived in the house in that condition for many, many years; in fact, the floors were never finished properly. What a wonderful place for varmints, scorpions, ants, snakes, rats, and who knows what else to enter and share our home with us.

For some sort of "protection," we got more cats to battle the rodent population. There were no window coverings except for a type of plastic matting hanging at each opening. Doors usually were just a cloth hanging from a rod at the top of the doorway. With the huge tropical rain, wind, and lightning storms that were a regular occurrence in that area, the mats would be blown out of the windows as we scurried to stay dry inside. Jansen Hall had one small front porch, with about five steps coming up from the ground below. The place looked like a palace compared to the mud, stick, and leaf houses. The Africans were no less than astounded!

From Mother's book, *No Certain Dwelling Place*:

"Coming to see the house the Africans were amazed with the brick two story house. Seeing the large rooms, table, and beds, they clapped their hands over their mouths saying, "*Aba, Aba*," an exclamation of surprise. Some of them refused to go upstairs, unable to believe the second floor would hold them. There were even two bathrooms, one for the boys and one for the girls. Each one had a square brick and cement bathtub, and a small tin sink. The toilet was also built of shaped cement. Because of the lack of plumbing for running water, there were two wooden platforms outside each of the two bathrooms. On each platform were three steel fifty gallon drums. African school girls from the "Fence" would carry water each day to pour into the steel drums so water would gravity feed into the sinks. Each girl, in true African fashion, would carry a huge tin tub of water on her head from a spring about a mile away, up a steep hill through the jungle. Many times up to half the water would spill as the girl carried it on her head, dousing her from head to toe. They cheerfully carried on."

My own memory of that is when my brother and I, with our bedroom right beside the water platform, would wake up and watch the African girls, with their clothes soaked to the skin, pouring water into the barrels. A few times we were caught by our parents and spanked hard. A person could draw water from the platform and then transfer it by hand to the toilet or to the bathtub. A similar platform was located outside the kitchen, for the same purpose of bringing water for cooking. The toilets drained through a pipe laid below the outside walls, into a pit in the ground. The sink and bathroom water did the same. The kitchen had its own drain.

Up a long homemade flight of narrow, steep wooden stairs there were three bedrooms and a large hallway. One bedroom was for the boys, a second larger one for the girls, and a third one for

Mother and Dad. Off each bedroom was a kind of attic area used for storage and a closet. In the boys' and girls' dorm rooms were wooden platforms with beds or bunk beds on them. All the beds had mosquito nets hanging over them. Some kids used the netting; others, like me and my brothers, did not. It became kind of a badge of courage to not use the mosquito net, as boys will do. Since we, the Lindquist boys, and the VanderPloeg boys were some of the first American boys in that region, we felt obliged to show some bravado now and then.

The boys' upstairs dorm room had one window. None of us wanted to go downstairs in the night to use the bathroom, so we fashioned a solution. Taking a foot-long piece of good-sized bamboo, we cut it to make a trough. You guessed it, we just urinated through that bamboo outside of our upstairs window, thinking it would drop harmlessly to the ground. One day my mother called a meeting of all the dorm boys outside below the upstairs window. Her question rang out like the clanging of a bell: "What on earth are those streaks down the bricks under your window and onto the ground!?" Mother, of course, knew the answer. All of us dorm boys had to do extra chores for a month for this indiscretion.

Of course, the girls would each use the bathroom before going to bed at night. We could lie in the boys' room and listen as they thundered down the stairs to the girls' bathroom. They would each be calling out, "First, second, third, fourth," and so on. Using this signal, sometimes we boys would sneak into the girls' bedroom and hide under their beds. The goal was to see what girls looked like as they changed clothing and went to bed. Well, well, one of us boys always giggled, and then the heat was on. Those were some of the worst spankings and chewing outs we ever got.

Around this time, a new missionary couple came to our station at Katanti. We really liked this new family and spent many hours with them. The father of the family had a paperback book called *Pogo*. One time he allowed me to borrow his book. Upon loaning it to me, he gave me strict instructions to take good care of it and to not loan it to anyone else. I read it daily with delight. One of the other boys in the dormitory where we lived asked if he could borrow the book. Ignoring the stern warning from the owner about taking good care of it, I foolishly loaned it out to my friend, with a warning for him to care for it gently. For some reason I don't recall, he got mad at me and chewed the paperback book into a beat-up chewed-up pulp. Oh my, now I was in trouble! To this day I have strong memories of Mother making me return the damaged, useless book to its owner next door. And I endured another long disciplinary speech, given by the disappointed owner. Additional punishment came from Mother.

Later, in Jansen Hall we received a wood-burning cook stove for the kitchen. Another of our jobs was to gather wood for the stove. It actually had a small oven, so we could bake bread, rolls, casseroles, and cake mixes. Our cook, a man named Kibekiangabo had really taken to his training by my mother and was a wonderful cook. He was our cook all the years we spent in the Congo. As far as I know, Kibe—as we called him for short—is still alive and in his nineties, although quite ill. I had no idea I would see him again in the last years of his life.

For clothes washing, we had hired two Africans. Their duties were to wash, dry, and iron the clothes for about fifteen of us. Early on, they used a pan and a washboard for washing. Then, as time passed, someone in America donated a Maytag washing machine, powered by a small gas engine not unlike a mower engine. The washer was the type that had a crank wringer. Our houseboys were all carefully trained in the use of these things. The gasoline available in the Congo often contained water, meaning the gas engines were always a problem, but they were better than washing by hand. For drying they used an old-fashioned clothesline between trees. To iron, we had brought from the U.S. some cast-iron old-timey irons in the shape of a small boat. One can see these today in antique shops around America. There was also a wooden/metal handle to hold and carry the iron. The irons were placed outside right into a firepit filled with red-hot coals. As the wash boy needed an iron, he would pick it up out of the coals by the handle, then wipe off the bottom with a rag, and then iron clothes. Often the surface upon which the garment was ironed was simply a table top. Later on, my brother Dave and other missionaries made wooden ironing boards.

Dad had a small office he called his study. In that space he could write sermons, Bible lessons, booklets, tracts, and serve as the mission's Field Secretary/Treasurer, as well. He also could have meetings with African teachers, evangelists, and elders in that room. It was the one room with a lockable wooden door, for his privacy. His desk was handmade from wood, the size of a large office desk with a drawer in the center and three drawers on each side. None of the drawers had locks on them. I must admit my brother and I were tempted occasionally to look through those drawers for things we might find interesting, or grownup secrets. Of course, we paid a price for that too.

Speaking of paying a price, that study was where our spankings were usually given. Dad had a habit of giving us a lecture before the actual spanking. Most of the time the lecture would go on and on, until we boys were ready to just say, "Please! Go ahead and spank us!"

One time Dad had many, many meetings over time with the African elders, discussing the tribal rights of passage. These included tribal facial markings, boys becoming men, girls going through puberty, polygamy, adultery, sex, and the like. They also told stories of the medicine men, witchdoctors, and the powers of evil spirits. The facts of these tribal rites were documented on paper in Dad's desk drawers. These disclosures to the white man, Dad, were also the subject of much heat, consternation, and trouble from the Africans.

In Balega culture these are sacred secrets, only known to those who have survived the rituals and who have become Balega tribal members. Revealing these sacred secrets to anyone outside of the tribe, especially white people, was punishable by death! A long line of Africans came to protest what they saw as deception by the African Christians for disclosing these sacred rituals. Objections came from witchdoctors, chiefs, elders, and regular folks alike. Dad was forced to make a solemn pact and promise that he would never, ever repeat these secrets to anyone in America, not even to his wife. So the heated calls for action were calmed.

However, my brother Dave and I had also come across those papers in the drawers. After much punishment, we were solemnly sworn to secrecy. I am proud to say that neither Dad, nor Dave, nor I have ever broken that trust we gave to the native Africans. Those papers were later destroyed in the turmoil of 1960. Undoubtedly, that was a good ending for them.

Whenever or wherever growing boys and girls are together in one house, inevitably boys will fall for girls, and vice versa. So it was in the Jansen Hall dorm. I have a picture of most of the dorm kids in front of Jansen Hall. My oldest brother, Eldon, is standing next to a high school girl who, at the time, was his girlfriend. As the photographer got ready to take the picture, Eldon's girl leaned over toward him as if leaning her head on his shoulder.

Also, on one outside corner of the brick house a brick had worked loose. We all decided that this would be the "mailbox" for guys and gals to write each other notes. Having written a note to a girl, the boy would place it in the brick hole, under the loose brick. Of course, throughout the day kids would "check the mailbox." When a note was found, great rounds of embarrassing teasing would break out toward the sender and receiver of the note. Of course, we secretly enjoyed the attention.

While we were living at Jansen Hall another boy/girl tradition was born. In the Congo, situated right on the equator, the full moon was a big deal every month. We kids would be allowed an extra hour to play outside, the moonlight being so brilliant. Someone had sent bicycles from America to some of the families. So each "couple" would take a turn with the boy riding the bike and his girlfriend on the handle bars, under the moonlight. The couple would ride about a mile down a long hill, then get off the bicycle and walk together, pushing the cycle back to the hilltop for the next couple. Of course, the walk back up the hill led to all kinds of shenanigans between the couple, as they came back up to the house in the dark.

Similarly, after dinner each night, all the kids and grownups in Jansen Hall would gather in the living room for a time of singing, reading, devotionals, and praying. Inevitably those with a girlfriend or boyfriend would sit together on the floor, chairs, or couch. And, of course, just like kids all over the world, the couples would hold hands. When we were asked to refrain from holding hands, we devised all kinds of ways to hold hands—under a magazine, between couch cushions, under a pillow, you get the idea.

My brother Dave and I found an old, used Whizzer motorbike someone had discarded behind the house, leaving it to rust. Dave, being quite a mechanic and handy with fixing things, suggested he and I work on it and try to get it running again. For a solid year, we spent hours and hours taking the Whizzer apart and putting it back together. The Whizzer was really just a bicycle with a motor over the pedal assembly and a gas tank behind the handle bars. From America we got some technical drawings and sketches, which gave us some guidance. And we ordered some parts from Europe. Finally, we got the Whizzer going again. We felt pretty cool being the only kids with a motorized bike. Never mind that it would not carry a load up the hill at Katanti, we enjoyed it anyway.

Our dining room had a Singer foot-pedaled sewing machine in one corner that Mother had brought all the way from Nebraska. She spent long, long hours mending and sewing on that little machine. She was quite a good seamstress, and made many of our clothes or mended others. Once, she opened the little drawer in the front of the sewing machine to get something she needed. Alarmed, she watched a long, thin, green snake uncoil and crawl out of its hiding place onto her lap. Quickly knocking the snake to the ground, she called out for help, and someone ran to cut off the snake's head before it could get away. T*his is Africa!*

From Dad's book:

"'Daddy, Mommy, help me! Come quickly!!' Patty, our little daughter, awakened us with her screaming in the middle of the night. Jumping out of bed, we ran downstairs to find Patty in the hallway between the two bathrooms, trembling from head to foot. One hand held the girls' bathroom door tightly shut. The other was holding a kerosene lantern. David was standing in front of the boys' bathroom door. He and Patty had come downstairs together to use the two bathrooms. He was trying to quiet Patty.

'Patty,' I asked, 'what is the matter?' Patty sobbed, 'There is a big black animal in the bathroom.' 'Oh, Honey,' I replied, 'there cannot be an animal in the bathroom. How would it get into the house? You must have been dreaming.' 'Yes, there is, Daddy,' she shouted. 'Take the lantern and see for yourself! It has big, bright eyes.' With that she ran to her mother. Taking the lantern, I cautiously opened the bathroom door. I really didn't think there was an animal in the bathroom. Then, I heard a vicious snarl. I saw two glaring eyes. Bang! I slammed the bathroom door shut and ran to get a broom. Coming back with the broom, I slowly opened the bathroom door and shut the door behind me. Grrrr, Grrrrr . . . there it was, the eyes glaring at me from behind the toilet seat, while continuing its vicious snarling.

Taking aim, I hit it over the head. It was dazed, but ran to the other side of the room. I continued to beat it over the head time after time. David ran out and returned with a baseball bat. Wham . . . wham . . . he hit it over the head, killing it. 'What is it?' I asked Dave. 'It looks like an *Idimu* (wild civet cat),' Dave replied. 'Well, it sure was vicious,' I said. I was as scared as Patty. Later, we discovered the wild animal had come in through our 'cat-door' which we had made with a hole in the wall of the house."

"Rog, wake up now!"

I was awakened by my brother Dave's anxious yell from the bunk below me. At this time, Dave and I were sleeping in a small bedroom under the stairs on the first floor of the dormitory, our home. Dave had heard an animal in the house, rustling and shuffling around the main floor. In the dead of night, we switched on a flashlight and went searching for the culprit. Suddenly we were greeted with a snarling, scratching, black animal that looked like a wild cat, perhaps a type of lynx or wolverine. It was fast and slick as can be, and there was no catching up with this one. After we woke up Dad, he armed himself with a long-handled broom, I grabbed my trusty razor-sharp machete, and Dave got a baseball bat. Round and round the living room and dining room we three chased the wild animal—Dave and I in our boxer shorts and Dad wearing a bathrobe.

Finally, we cornered the intruder in the dining room. However, he was so fast he could go from the floor to the ceiling and back in a flash as he tried to stay ahead of the three of us trying to kill him. On one pass Dad hit him with the long-handled broom and Dave hit him with the bat. The two blows split his back open, and he began to bleed profusely from the wound. Going insane with pain and fear, the creature increased the quickness of his movements, trying also to attack us, his attackers. With a stroke of luck, one of us was able to pin him behind a cabinet. Blood was spewing everywhere. I was then able to strike him hard with my machete, cutting deeply into his neck. The neck wound was fatal, and we watched him die in front of our eyes. With the danger over, we all agreed that there would be no more sleep that night.

As the seasons changed from the dry season to the rainy season, it was time to re-thatch the roofing on the mud, stick, and thatch houses.

From Mother's book:

"The head evangelist was reroofing his house. Our own boys, Eldon, David, and Roger, were up on the roof with razor sharp machetes helping cut the vines which held the old leaves to the rafter poles. Suddenly, Eldon cried out in pain. Roger had been cutting alongside Eldon, swinging his machete in rhythm with Eldon's own cutting. Somehow they got out of sync. Eldon reached to pull on the vine, and Roger's machete came down across Eldon's two fingers, the index finger and the long finger. The blow almost severed the left index finger, which was dangling just from a bloody shred of skin. Two Africans carried Eldon to the dispensary for treatment. Another African ran to call for me. I had sewn on fingers and toes many times, but I looked for a way to take care of Eldon without using stitches. Eldon was already a good piano player, and I was extremely worried about the effects of this injury on his ability to play the piano again. Cleansing the ends of the stumps, I pushed them tightly together, wrapping them snugly and tightly to a tongue depressor. Having no idea if this would be successful, we all prayed diligently for good healing. Praise God, He took over and the finger(s) healed perfectly in place. Eldon's ability to play the piano and organ had been saved!"

Life rolled along in Jansen Hall, the dormitory for the Berean school for missionary kids. At the dorm we had a monthly game night, or sometimes game weekend. We would either play board games or sometimes have what was called back then a "Blind Box Dinner Date." In that case, all the girls' names would be put into a container together. Each girl would also make a box dinner for a picnic for the night's activities. Then the boys would do a blind draw of each girl's name from the hat. Whatever girl's name was drawn became that boy's date for the boxed dinner and evening date night or games weekend. Although there were some shenanigans to try to rig the system, most often it worked fine. It included many "surprise" couples for one time together. Sometimes we would include softball games, badminton games, and tennis games. In Jansen Hall the dining table was just the right length for a table tennis table, although it was a bit too narrow. No problem. We played many, many games and tournaments on that handmade wooden table. By the end of "games weekend" there would be crowned a champion in each of the sporting events.

Many years later, we Green brothers learned of another name coined for us by the missionary kids: "The Legends." I suppose it was because we were among the first group of missionary kids who came to the Congo. In any case, we wore it well.

Behind Jansen Hall were a number of guava trees and fruit trees we had planted, which were now a good size. One day we three Green brothers decided to each climb a different tall tree in the backyard garden. From the top of the three adjacent trees, someone gave a signal and my brothers and I sang the hymn "Victory In Jesus," belted out in three harmonies from the treetops. To this day some of the kids from Berean Christian Academy school remember the singing from the trees, by the Green boys.

CHAPTER EIGHT
CRAZY STUFF!

ONE OF THE key phrases in the Congo in the 1950s, which went right along with "This is Africa!" was this: "Anything that can happen, will happen." Conditions in the country at that time were quite unpredictable. So we faced each day as an adventure with a capital A.

A couple of years after we arrived in the Congo, my little sister, Patty, learned how to walk by herself. She took great advantage of that and could be found almost anywhere at any time. Also, during those early years there were many monkeys making their homes in the trees around our houses. One day I heard a shriek from the space in front of our house. Patty was standing there holding an empty banana peel in one hand and screaming. On top of her little head sat a monkey, calmly eating the banana he had stolen from Patty.

We boys were also treated to the discovery of what the native Africans call a *kipoi*. Here's how it worked: Two African boys stood a few paces apart, holding two long poles, one on each of their shoulders. In the center of the poles was constructed a wicker hanging chair, attached to the poles. The Africans knelt down, and the person to be carried climbed into the chair. The guys stood back up and then carried the passenger wherever they desired to go. I must admit we took some advantage of this way of moving about, until our parents stepped in to stop us overusing this kipoi as a free taxi service.

On a serious note, one dark, moonless night, one of the first six missionaries, Mamie Jansen, got word that someone needed her down in the Africans' village. Since her husband wasn't available, she decided to walk the mile down the hill by herself to help the Africans. As always, she took along a flashlight. No one ever went anywhere at night in the Congo without a flashlight. After Mamie had taken care of the need in the village, she said good night and started her slow climb back up the path and through the trees to the home she shared with her husband, Albert. As this was a mile return walk uphill, it took her quite a while to reach her destination. She later related how every so often she would hear something outside of the circle of light from her flashlight. Each time it seemed to be behind her. But each time, she couldn't see anything in the darkness. So, breathing a prayer, she would press onward. This happened again and again, numerous times, in fact. Finally, she reached her house and went safely inside.

At first light the next morning, the Africans came running and shouting over one another to see if Mamie was hurt. There, at each place where she had paused to look for the source of the noises, were paw prints and tracks from a leopard that was stalking her.

How they thanked God for His protection. Kind of like when God shut the lions' mouths for Daniel in the lion's den in the Old Testament. This, too, is Africa!

Some of my favorite things to do were to go bird and squirrel hunting with a homemade slingshot and a bag full of stones for ammo. From an early age, I became a "crack" shot with the slingshot. I could even take a hummingbird, or at night a bat, out of the air with it. We cut a piece of a forked wooden limb from a tree, which served as the slingshot frame itself. Then we used a razor blade to cut long rubber strips from used tire tubes. The strips would be wound around and secured to the top of the forked stick. At the other end of the rubber strips we would cut out a leather pouch from the tongue leather of an old shoe. (Okay, okay, sometimes we cut out the leather tongue from a new shoe!) This pouch would hold the ammo, a smooth stone picked up from the ground.

My friend Tim and I would wake up early nearly every day, around 5:00 a.m. We would meet and spend an hour looking for smooth stones for our slingshots. Around forty-five minutes later we would be joined by Yoanne and Yosef, my two best African friends our age. Off we'd go "*Ku Bisindi*," which in the local dialect meant "to hunt squirrels," walking or hacking our way through the jungle. (To this day, if you were to ask Yoanne or Yosef, "Where's Roger (*Banuamazi*)?" they would answer "Ku Bisindi.") Almost every day we made at least one kill and brought it back to the mission station. Usually the dead squirrel was carried in the pocket of my shorts, bleeding down my leg. Once in a while, I'd take it to school and gross out the girls. Our teacher (Mother) was not impressed.

These squirrels and chipmunks had hardly any good, edible meat on them. Although once in a while we would skin and cook them, there wasn't enough meat to really make a meal. So Yoanne and Yosef would take them home with them.

Also, we learned from the Africans how to make traps for the same types of animals in the forest. We used a bent sapling tied with a vine to a circular trip wire, and the chipmunk or squirrel would run along a horizontal branch into the trip wire. The sapling would spring back upright, tightening the trip wire and noose around the small animal. The craziest thing I remember catching was just the two front teeth of a chipmunk! The rest of him was gone, minus his two front teeth.

The cars and trucks owned by the mission would need new batteries from time to time. We kids found a great use for the old discarded and no longer useable car or truck batteries. We would melt the lead plates inside the battery. Then we would form the molten lead into round pellets for even better ammo for our slingshots.

The students from the Berean mission kids' school worked together on several sports projects. On level ground across from our schoolhouse we cleared and built a softball field. Bats, gloves, balls, and other equipment were sent from the U.S., and highly prized. The field was lined with bleach for base paths and had wooden bases and a pitching rubber. It had a short left field where many home runs landed on the roof of the church building. Center field and right field had a road crossing right through the middle of the outfield, with a pretty good downslope into deep right field. We did the best we could with what we had.

Each semester, the entire school would be divided up into two co-ed teams. Boys and girls alike would play games every

Thursday and Friday throughout the season. Umpires were usually one or two of the mission men, or we made it work without umpires. At season's end, we held a big playoff to determine that season's "World Series" winner. Great fun! (By the way, even in the 1950s, we could listen to the real World Series via shortwave radio, on Armed Forces Network.)

A much bigger project was the construction of a regulation-sized clay tennis court for our use. The land we picked was below the schoolhouse and had at least a forty-five-degree downward slope to it. Carefully, we measured the size of an official tennis court. Then the excavation began, with us kids hand digging soil from the upside of the slope and carrying it to the downside. My recollection is that it took us about a year to complete the excavation and earth-moving, all done with hand tools by us kids from the mission school.

Once in a while we'd talk our African friends into helping us. After the excavation and packing of the clay, we built a tower on one side for the head linesman, and fences on three sides to corral stray balls. The high side of the hill served that purpose on the fourth side. The clay was lined and a net was sent from America along with tennis balls and rackets. In those days the rackets were all wooden. The finished court could be used for tennis, basketball with goals, badminton by raising the net, or volleyball. Again, we set up brackets, tournaments, and competed fiercely for "top dog" status. Often, the finals would come down to myself against a girl my age. I must admit she usually won the finals. Linda Hendry Coolen was hard to beat, at any sport. We also had doubles brackets, which she and I sometimes played together.

The national sport in Africa is soccer. Over there it is called "*futbol*" and is wildly popular. Even more popular than American football, basketball, or baseball. The African schoolboys had a regulation-sized soccer field near the African village. Their school had a team, which would play other regional schools.

The soccer balls were made of very hard leather, and the Africans always played barefoot. The players were very competitive. Getting hit by one of those hard leather balls would certainly leave a mark. I asked if I could play on the African team, barefoot of course. After much debate and negotiating, it was decided that I could play goalie, with no gloves, and yes, barefoot. The Africans in that decade were very afraid that I'd be injured and they would be the cause. In those days that would have resulted in punishment for the African perpetrator. After convincing them that I would not call for punishment should I be injured, I was allowed to play goalie with the team. Later, I was allowed to play the field with the team.

One of my proudest moments was being carried off the field by my African teammates after playing goalie for a big regional game. The opposing team had a guy with a huge leg who was determined to score against the white guy, me. The game against the arch rival ended in our favor—a rousing 3-0 win!

CHAPTER NINE

CRAZY!—AND NOT SO CRAZY

WE SPENT QUITE a bit of time trying to rid ourselves of a small bug we called a "jigger." I'm sure that was not the scientific name, but it came after us with a vengeance, especially those of us who were prone to going barefoot. This little devil would burrow in along toenails or fingernails, making a pouch filled with fluid, where it would lay its eggs. When the eggs hatched, the pouch would break, sending fresh little guys all over the toe or finger to dig in again. The effect of these little dudes was crazy, crazy itching. So the best way to get rid of them was to soak the affected area in hot water, to soften it up. Then one would take a sharp needle and carefully dig around the fluid pouch filled with eggs. Care had to be taken not to break the pouch. After much hard, careful work the whole pouch could be extracted and then burned on the point of the needle. Finally, we would soak the toe in iodine, or Absorbine, Jr. Alas, these guys were so persistent, the whole procedure had to be repeated many, many times, often every night.

A very interesting day as recorded in Dad's book:

"Hearing strange noises I asked the African preacher what was happening. 'They are having a healing meeting,' he answered. Missionary John and I wanted to see the antics of the witchdoctor in charge of this "healing." Stooping down, we worked our way into an open shed in the middle of the village. Roger and David, who had come along with us, did the same. In the middle of the shed was a fire for light. People were crowded all around, watching the witchdoctor work. Not quite in the center lay the sick man, in the arms of another person. There beside him was the witchdoctor, a fearful sight! He wore a special hat and leopard skin around his waist. His face was painted grotesquely and he held rattles in his hands. Behind him four drummers were keeping up a constant beat on skin-covered drums, as the medicine man rubbed the sick man with his concoction and blew into his eyes and ears. Suddenly, he grabbed his rattles and started to dance wildly about. Shaking, jumping, jerking, and flailing about he sprayed chicken blood all over the man, the crowd, us, and the building. His garbled grunts, groans, and shouts were scary, but unintelligible. At once he stopped in front of us, glaring menacingly at us. He said, 'There, white men, if you've come to see it, you've seen it!' To say the least, it was quite scary. Finally, his contortions ended with a flourish.

We walked to the other side of the building to shake hands with someone we knew. Turning, I reached out my hand to the witchdoctor. Not knowing what to expect, surprisingly he shook my hand and smiled. 'You wouldn't allow me to speak a little bit?' I asked hesitantly, in African fashion. 'Speak on,' he said. I began to speak. 'Yes, this man is sick and really needs help. But, what you are doing is worthless. In fact, you ought to take him to a hospital where a doctor and nurses could help him.' Looking around, I didn't see anyone getting angry, so I continued to share the Truth of the Gospel, and the healing power of the 'Great Physician.' Finishing I said, 'I want to pray for this sick man.' Bowing my head, I did so. Shaking hands with the witchdoctor again, and thanking him for letting me 'speak on,' we went to our car."

Author's note: This scenario became a means to share the Gospel of God's love over and over. Dad would make a deal with the witchdoctor(s), that they could do their "thing" if Dad could "speak on" after they were finished. Amazingly, that deal worked time after time.

One of the problems, especially in the early days, was the fact that the Africans were dead serious about worshipping idols and spirits. Almost every home had a small shrine, either indoors or outside, to house the many spirits the Africans worshipped. Sacrifices were regularly made to these idols of wood, stone, or ivory. Sometimes a small house, replicating the family's own house, was built right beside the family's. That house was to house the evil spirit. In addition, they believed in many curses of evil spirits. The witchdoctors perpetuated these beliefs to their own benefit. One of the spirits was called Kimbilikiti. People believed he or it was "as skinny as a bicycle, and as big as a house, and he could move right through walls." It was also said, "If you leave out a plate of food when you retire at night, he will eat it and give you favor."

Old ways die hard and are hard to change. Many times people would wait to bring their loved ones to the dispensary or clinic until many strange evil practices were tried first. Then, when they finally came to the hospital, the medicines would be administered too late. Years of teaching, preaching, and hard work by the missionaries accumulated to teach the Africans not to trust in evil spirits, practices, and witchdoctors, but to trust in the one true God.

One day my African friends were going hunting in the jungle. My friend Tim and I asked our parents if we could go along with them. After receiving permission we were excited to go. This time we planned to go farther into the jungle than we had ever gone. In fact, we would be over the top of the farthest mountains visible from our homes. We would indeed be far out of sight of our mission compound. Our message from our parents was to be very careful going into these new surroundings. As for me, I was jumping up and down with excitement. After all, the Africans said that to go that deep into the jungle would be going where dangerous leopards, wild boars, elephants, hyenas, water buffalo, chimpanzees, gorillas, and other predators lived. What could be more exciting to a boy my age? Grabbing my trusty slingshot and tucking a machete into my belt, I raced out of the house, shouting, "*Twendazi*. Let's go!"

Off we went deep into the jungle, where the native Africans had a trap line set for little and big game alike. Laughing, hollering, and yelling, we were all having a great time. Suddenly, I took off on the run ahead of the others. Rounding a corner and slowing down to a walk, I saw a six-foot-wide clearing in the jungle, right in front of me. While it looked slightly different due to the fact that the jungle was cleared away, it was in the path in front of me, with leaves and limbs lying on the ground. Giving it only a second look, I stepped into the clearing to the shouts of the Africans. "Don't go through there!" As I continued forward, suddenly, with the swishing sound of breaking sticks and leaves, I was in free fall into a deep, dark pit!!

"My companions, I'm dying!" I yelled as I hit the bottom of the pit, about thirty feet down. Racing up, my friends were frantic with concern. They knew the clearing was a covering over a deep pit, dug to catch bigger game walking down the trail. The Africans also knew the custom was to pound pointed poison-covered wooden posts into the bottom of the pit, with the tips pointed upward—to ensure the death of the hapless animal. Fearfully looking into the pit, they shouted, "*Banuamazi* (Roger), are you alive?"

Having dusted myself off and determined I had no broken bones, I said, "Yes, I'm okay, but get me out of here."

Grabbing some forest vine, they made a strong rope and lowered the vine into the pit. "Put it under your arms and around your front," they yelled. "Hang on tight, and we will pull you up and out." Pulling together, they lifted me out of the pit. "Are you all right?" they asked. Although I was scared out of my wits, I was all right. Only some cuts, bruises, and pains. We thanked God this was a *new* pit trap and the owner had not yet put poison stakes in the bottom.

From Mother's book:

"'Help me, help me!' The cry came from a woman who was taking a bath and washing clothes at the side of the Kasagi river, about five kilometers from the mission compound. Running quickly to the spot, the people saw that a huge crocodile had swum up behind the woman, opened its giant jaws and clamped its mouth right around her hip!

Yelling, kicking, and screaming the woman grabbed a small tree near the river's edge with both hands. She was holding on for her life. Then, with her free leg and foot she kicked the crocodile in the eye. But the crocodile would not let her go. African crocs almost always grab their prey, swim into the water and then dive in what is called the 'death roll.' Once that happens it is too late for the unlucky prey. However, before that could happen, someone came running up with a spear and jabbed the croc in the head and mouth. Ripping and tearing huge chunks of flesh out of the hip, the croc let loose of the woman and dove into the depths of the river. As it dropped out of sight a group of men came running to carry her the five kilometers to the mission clinic.

Such ugly teeth marks, such pain, such gaping wounds. I tried my best to patch up and sew her many, many wounds—and save her life. She had lost a lot of blood. I said to her family, 'She can praise the Lord that she is still alive.' To our great joy, the woman lived, and was even able to walk again."

The jungles of Maniema and Equatorial provinces in the Congo rainforest are crisscrossed with good-sized rivers. These are used for transportation, nearly always by means of huge dugout canoes. With the coming of the white man with autos and trucks, a need arose for cars and trucks to cross these rivers. Bicycles could be put into a canoe to cross, but vehicles were a much bigger issue.

The Africans devised an answer to the problem. For some unknown reason the answer was called a *bac*, the African word for ferry. Three or four huge canoes were the bottom of the bac. On top of that, a wooden platform was lashed to the canoes with very strong vines from the jungle, almost as strong as steel bands. Then, three planks were cut and placed on top of the frame, to act as two drive paths for the vehicles. Primitive on and off ramps were made from jungle materials and slid into place on each side of the river as necessary. The final piece was a good-sized steel cable, imported from Belgium, which was hung across the river and attached to a huge tree on each side of the span. Along the cable was a rolling cable device, also from Belgium. From the roller was a cable leading down and attaching to the bac itself. This would keep the whole thing from going downstream in the current. African men would then use long poles to push the ferry and its load across the river.

When one would drive up to the river, the first thing was to see if the ferry was on your side of the river. If not, one would lean on his vehicle horn for a long, loud blast. *If* the boat driver was there, and if he was not asleep, he would bring the ferry over to the side where the vehicle waited.

Many were the hours and hours we waited under the tropical sun for the guy to awake, or come back from wherever he had gone. We also waited many hours at night. One had to be extra careful at night, as the crocodiles would leave the river and hunt for prey, human or otherwise, on the riverbanks. Once again, *This is Africa!*

Africa does not work on the same concept of time as the Western world. The African pace is quite slow, and time really doesn't mean much at all. For example, if someone told you they would come to your house at 9:00 a.m., that really meant any time between 9:00 a.m. and 3:00 p.m. If the church bell (an old steel truck rim) was rung at 9:00 a.m., the Africans just then started to get ready to go to church. They never planned ahead, based upon time. Africans do not understand why Americans and Westerners are so driven by time. Seriously, they have much less stress and worry than Americans.

CHAPTER TEN
LEARNING THE LANGUAGE

AS SOON AS we settled in the Congo, the first order of business was to learn the language(s). The local tribal language was *Kilega*, spoken everywhere across our mission region. The "trade" or business language was *Kingwana*, a version of Swahili. This was used in many towns, especially in trading or negotiating. The Belgians started schools to teach the Africans French—the Belgian variety—which was to become the Congo's national language. To this day, many in government and schools speak French.

My parents were taught a daily lesson of Kilega by Mamie Jansen, one of the original Berean missionaries. She would give them grammar and writing lessons each day until around 1:00 p.m. Then, after the obligatory Congolese "one-hour siesta," my parents would go down to the native village to practice speaking and hearing what they learned that day. Many, many stories came out of these learning and practicing sessions, far too many to tell.

After quite a long time, my parents learned Kilega very well. Kingwana, not so much. That came later, when they moved to one of the towns where Kingwana was used much more often. As for me and the other missionary kids, we did not take any lessons in the local dialect. At grade-school age we simply went out each day and played with the Africans. After just a month, we knew their language very well. (Interestingly, the Africans did not learn English from being with us). We kids were often a source of irritation to our parents regarding language study. They had formal lessons, and we learned quicker and better with no lessons. Many times our parents would ask us to interpret something the Africans were saying to them.

At one point Dad said: "I might as well give up studying the language and just go play with the boys; I might learn it sooner". In fact, the Africans gave Roger one of their greatest compliments: '*Banuamazi* (Roger), speaks our language of Kilega just as well or better than we do!'" (That theory would be tested far into the future.)

We three boys were not allowed to go near the house during the times that Mother and Dad were studying the language. Well, we didn't always pay attention to that rule. Daily we would stand outside the window of the teaching room at noon and become a distraction while waiting for our lunch. Soon we were banished to the porch of our own house a distance away. Undaunted, we'd stand on our porch at lunchtime and sing at the top of our voices: "Here we sit like birds in the wilderness, waiting for our food." I'm afraid the effect was the same: banishment.

One day we heard lots of yelling and commotion from the direction of the dispensary and clinic. Since Mother was already teaching us in school, an African came running quickly to get her help. Upon arriving at the clinic, she saw an African lady who had been carried in on an African *kipoi*, the African chair built to carry someone. It was about noon, and very hot.

"What's the problem?" Mother asked.

"She's been bitten by a poisonous snake," answered the men who carried her, wiping the sweat from their brows.

Seeing the woman looked very sick, Mother asked, "Why didn't you bring her quickly here?"

"We used our African wisdom on her first" came the weak reply. The "wisdom of the Africans" was this: If the snake itself was killed, its head was cut off, boiled, and then some of the boiled broth was given to the victim to drink. If the snake got away, the Africans called one of their witchdoctors who (supposedly) was "wise in snake lore." The medicine man or woman would then spit on his or her finger and hold it over the place where the victim had been bitten. That was it. (Yes, for this they charged a fee of a chicken or two!) Frustrated beyond words, Mother gave the lady a quick shot of antivenom for this particular snakebite. Then everyone sat down to wait and pray. Sadly, too much time had been wasted with African voodoo, and the lady died and was carried away for burial. Life in Africa was very dangerous.

One evening just as we sat down for dinner, we saw many, many Africans running past our house toward the neighboring missionary house. They were all screaming and shouting and seemed to be out of control. Dad told me to run to the door and find out what was happening. As I looked out, I saw a horrible sight. The neighboring missionary house, where my friend Tim lived with his family, was on fire! The roof was totally engulfed. Africans and missionaries ran about, futilely trying to dump water from buckets onto the flaming roof. Thank God the entire family of seven was able to escape out the doors. Try as they did, the meager efforts were useless in fighting a fire feeding itself on dried sticks and dried thatch for roofing. The house and all their belongings were burned to the ground. Missionaries and Africans alike stood crying as the house collapsed. The family had only what they had run with to the outside.

The fire had been started by an old-style Aladdin kerosene lamp with a very tall chimney, one that was too close to the thatch roof. Comforting them as best we could, that night we took Tim and his family into our home just next door. The next morning Tim's dad told my dad how awful it was to wake up and realize you had nothing. Grateful that there was another smaller house available, the next morning we immediately moved our family to a smaller house, as our number was smaller than Tim's family. I vividly remember that first morning after the fire. Tim and I were digging around in the ashes of their home, looking for just anything. The coals were still hot, and upon picking up a penny I severely burned my hand. Of course, this was nothing compared to their loss.

Yes, you've read many stories, some crazy, and some even crazier. But none of us knew the crazy, and even crazier, extent of danger to which our beloved Congo was headed in the future.

CHAPTER ELEVEN
TRAVELING TO AMERICA

FIVE YEARS INTO our first tour in the Congo, in 1955, it was time to return to America for a one-year furlough from the mission field. We left with mixed feelings. After all, the Africans had become our closest friends and we would miss them. On the other hand, we missed our extended families in the U.S. It was also time to touch base with the many, many churches and wonderful people in America whose continuing donations and financial support made it possible for us to minister in the Congo. So, it was time to go.

On our flight(s) from the Congo to New York City we traveled on four-engine propeller airplanes for the very first time, on Sabena Airlines and Pan American Airlines. What fun that was! In the darkness of nighttime, one could look into the engine cowlings and see the red-hot parts of the engine just humming along. In those days, the four huge propeller engines caused significant vibrations, as well. The service, the meals, the flight crew, and the beautiful young flight attendants were a sight for us to behold. Meals were served on white cloth-covered trays, and almost any service was available. Even the meals tasted great. Many people dressed in suits and dresses when flying during those times.

Our route of flight took us from Bukavu, Congo, to Bujumbura, Burundi, to Leopoldville, Congo, and across the ocean to the Azores Islands for a stop for fuel. From the Azores we flew to New York City. In those days, there were no jet bridges, so we unloaded down the plane's front and rear staircases right onto the tarmac.

We had boarded in tropical Africa, wearing only T-shirts and shorts. The wind across JFK airport cut through us like a knife. Wow, welcome to America. Upon clearing customs we were met by one of Dad's seminary friends. After one night, we headed cross-country to St. Louis and on a trip to visit relatives scattered across the U.S., all the way to Oregon and California.

One of our first stops was at a Howard Johnson hotel and restaurant somewhere in the eastern U.S. The chief purpose of the stop was to buy a delicacy we had not had in over five years, a wonderful ice cream cone. If memory serves me correctly, it was five cents.

Our travels over, we settled down on a chicken farm just outside of Omaha, owned and operated by my uncle and aunt on Dad's side. These dear relatives had opened the entire second floor of their farmhouse for us to live for the year. Another of Dad's sisters and her husband and our cousin lived just a half hour away, in the city of Omaha. This gave us a great chance to get to know our relatives.

We kids attended Florence Grade School in North Omaha, while Dad took many, many trips to churches all across the country. On weekends and school breaks we would often travel with Dad. We three kids played music in a family group as we traveled to churches and told of the mission work in the Congo. Eldon played the piano, Dave the accordion or trombone, and I played the trumpet. Patty would sit with Mother in the very front row. Dad would give a short talk and tell a story from his book, *Uncle Ernie's African Stores*, for the kids in the crowd. Next was a demonstration of African drums, then he would show slides of Africa and the Congo. Dave and I usually worked the slide projector.

Also while in the U.S. we spent part of the summer on the cattle farm of Mother's dad's old home place in northwest Nebraska. On that farm, we three boys really learned how to work, and how to drive a tractor. Dave and I found an old Model T Ford in disrepair in the trees behind the farmhouse. We spent many joyful hours trying to repair and drive the old car.

Author's note: On a return trip there in 2014, I was able to retrieve several pieces of rusted steel from the body of that old Model T.

CHAPTER TWELVE

CONGO—THE EYE OF THE STORM!

SOON OUR FAMILY'S furlough would be over. We needed a good number of fifty-five-gallon barrels to pack our things into for the long trip back overseas. We found these barrels at bakeries, where they had been used to ship flour and sugar. Packing for a family of six was no small feat. In addition, someone had donated a small upright piano to our family. We shipped it ahead to St. Louis, to be crated carefully for the overseas trip. Our family again made the rounds of relatives, friends, and supporters, complete with many sad goodbyes. Along the way we drove into Canada for the very first time and saw the amazing Niagara Falls.

When we made a trip to a friend's Ford dealership in a small northwestern Nebraska town called Chambers, the friend arranged for us to buy a used 1956 Ford Country Squire station wagon with three rows of seats. Someone made a large wooden box to put on the car's roof, for all our stuff. Many nights, Dad, Mother, and Patty would sleep in a motel or someone's home, and we three boys would hang curtains, put down the third-row seat, and sleep in the back of the wagon. We used that car for the next three months, driving to churches from the west coast to the east coast before embarking on our second ocean trip.

Arriving again at Dad's friend's house in New Jersey, we discovered our ship was to sail in just two days. So, after a flurry of handling last-minute details, we, our car, and our belongings were loaded aboard a ship headed across the North Atlantic to Antwerp, Belgium. I cannot remember the name of the ship.

When we left New York Harbor at 9:10 p.m., it was exactly the same time and the same Pier fourteen from which we had sailed away six years earlier on our first voyage to the Congo. This time another missionary couple was going with us. We stood on deck, singing hymns as we passed the Statue of Liberty. Then Mother, the others, and Patty went to bed. We three boys and Dad stayed on deck into the wee hours of the morning, until the last lights of America were no longer visible. Our North Atlantic seven-day crossing was calm the first three days, but the next four were very rough. We were on the busy shipping course. Once again, we were going in a freighter, which had ten or twelve passenger cabins. We three boys stayed in one cabin, while Mother, Dad, and Patty had an adjoining cabin.

I vividly remember the night of the big storm. Huge waves, seeming like mountains of water, buffeted the ship and broke over the bow. Sometimes the same would happen at the stern. Terrified, we three boys prayed in our room and then, as kids will do, planned our escape if need be. Each of us slept in our T-shirt and underwear, and put our shirt and pants hanging right next to us on a chair. The idea was that we would be able to jump into our clothes and run to help. Thank God those plans were never needed.

We were happy when we first entered the English Channel and began to see fishing boats. Passing the White Cliffs of Dover, we turned up the River Scheldt and got in line for the locks that would raise our ship to the level of the harbor in Antwerp, Belgium. Waiting in the line of ships, it was damp, cold, and foggy. Dad and us boys stayed on deck through the locks and the docking at Antwerp.

My parents had decided to return to Africa via Belgium so that Mother could learn French and take an Infectious Diseases course and test in Brussels. The test was only administered in French. In order for her to be certified to do certain medical procedures upon her return to Congo, she would need the certificate from this course. So the plan was that Patty would stay in Belgium with my parents, while we boys flew on to the Congo alone. While we were scared of traveling internationally alone, we were more than ready to get back to the Congo, and our schoolwork. So, praying for our safety, and with some misgivings, our parents put us on Sabena Airlines to the Congo. Eldon was fourteen, David was twelve, and I was ten.

At the airport we discovered a missionary who was to be on the same flight with us to Africa. While we didn't know him, it was nice to know there was an adult to whom we could turn, if necessary. Thankfully, we had no problems.

This time our route took us through Athens, Greece; Cairo, Egypt; and Leopoldville, Congo; then on to Bukavu, Congo. There our dear missionary friends the Kennedy family met us for the overland trip back to our mission compound at Katanti, Congo, Africa. Remember, in those days there were no international phones, satellite phones, Internet, or even telegraph to communicate from the jungle to our parents that we had made it safely. They just had to take it on faith.

We boys lived with the Kennedy family for the six months that Mother, Dad, and Patty were in Belgium. My parents and sister had time for a quick three-day trip to the Alps before leaving Belgium for Africa. On their last night in the Alps they took a train from Lausanne, Switzerland, to Belgium, and onto a plane for the Congo. From Leopoldville, after rising at 4:00 a.m., they took a flight to the Kamembe airport in Rwanda, just eight kilometers from the Congo border town of Bukavu.

Author's note: The Kamembe, Rwanda, airport is still in use today. I visited there more than fifty years later.

From Mother's book:

"Two days later we were on our way by car over the same old 'Route de Kimbili' toward our boys at Katanti, Congo. Most of the trip was through rain and mud. Reaching the 'Katangila,' a clearing from which one could see the rooftops of our hometown mission post, we strained for a sight of the mission compound. Suddenly we were going up the hill on the driveway to Jansen Hall. Africans lined the road on both sides, waving, singing, and smiling. Many were carrying palm fronds, even making an archway over the road. Stopping in front of the house, our own three sons crowded the car doors to greet us. What a joyful reunion! We talked and talked half the night. We were back in our African home, and together as a family again."

None of us could have known that some fifty-seven years later I would have the amazing experience of riding up that same drive, returning again "home"!

CHAPTER THIRTEEN
DANGER AHEAD!

BEING BACK IN the Congo, at our Berean school in Jansen Hall, our home, and with our African brothers and sisters, was terrific. Our mission kids' schoolhouse had been replaced with one made with cement blocks and a corrugated tin roof. A front porch had been added to Jansen Hall, as well. We walked around the mission station the first couple of days, seeing what had changed while we were in America. We were pleasantly surprised, as progress had been made.

Fruit trees, pineapple, guava trees, avocado trees, and many others that we had planted in our first tour of duty had grown and were bearing fruit profusely. We kids in the mission dormitory spent hours and hours climbing in those trees, and also in the bamboo that had sprung up "volunteer" on the compound. A tradition was born among us boys and girls. We would climb up into a guava or citrus tree to carve a heart in the bark of the tree. Within the carved heart we would carve our initials plus the initials of our girlfriend, there to remain for posterity. Example: *RG + CW*, with a heart around it. Many, many years after the fact, the urge to do this would strike me again.

Yoanne and Yosef, my African best buddies, had grown just as we had. They marveled at our stories from America, while we listened to their tales of the Congo, especially the rumors of coming changes. It was such a pleasure to reconnect with them.

We started a field hockey league, to be played on our softball diamond. We trained the Africans in both baseball and field hockey during this time. For hockey sticks we found a tree in the forest called the umbrella tree, for its unique spreading branches. It also had a unique root system. About ten feet from the bottom of the tree trunk a root would grow out of the tree trunk at a right angle, parallel to the ground. After about twelve to fifteen inches, the root would again make a sharp downward turn and then grow into the ground. We would cut the root right against the trunk at the root's top, then we would cut the bottom of the root off just above the ground. The result would give us a great crooked stick that could be honed into a hockey stick. For the field hockey ball, we simply used a softball. We had great times and heated competitions in those field hockey games.

At that time, all of us mission kids were either going through puberty or approaching puberty. This was one of the big changes in the missionary kids, and our school for them, as well. Of course, our parents had this new thing to be aware of with the kids. Any number of times we kids got ideas to do things we shouldn't have done. Consequences were quick and severe. In addition, a very qualified couple, the Hendrys, had come from the state of Texas to take over the teaching at the Berean missionary kids' school. All of us kids liked them very much, and this change gave Mother a relief from that duty. This was good for her, since she could devote more time to the dispensary and clinic.

When the couple who had taken charge of the dormitory while we were in America moved to another place, Mother and Dad moved our family back into Jansen Hall, with the rest of the schoolkids. My parents' routine became much like it had been before leaving on furlough, teaching in the Bible school and evangelism for Dad, and running the dorm and helping in the dispensary for Mother.

A constant struggle in Jansen Hall was the battle against mice and rats. The varmints had discovered that in the attic space there were many barrels of rice and peanuts. Each time someone would remove some of these staples for cooking, they would spill some onto the floor. On any given night, we could hear the scurry of little feet running to the barrels to pick up the dropped peanuts and rice. Occasionally a mouse or even a large rat would fall into an empty barrel. We would spend the rest of the night hearing the critter trying to jump out of the fifty-five-gallon barrel. Time and again it would fall back to the bottom of the empty barrel with a loud bang. This went on for hours as the little guy tried desperately to save himself. At long last we got two more cats, to fight the rodent population.

CHAPTER FOURTEEN
THE WINDS OF CHANGE

THROUGHOUT THE LATTER half of the 1950s, strong winds of change began to sweep across Africa as a continent, and the Belgian Congo by extension. Revolts and calls for independence from the foreign colonizers of Africa ran rampant. Colonizers, including France, Belgium, Great Britain, Portugal, and others each reacted differently as the movement began to turn more violent. As for the Belgian Congo, its independence came a bit later in these events. By 1959–1960, the Congo was like a simmering volcano, ready to erupt politically at any time. The white missionaries, Belgian government officials, ex-patriot traders, and missionaries of other faiths were caught in the middle of the fast-moving uprising.

To the Congolese, "independence" meant no longer having to work, at all. No longer having to go to school. No longer being asked to work for a day's wage or for food. It also meant the native Africans would take charge, and the outsiders, including the missionaries, would be subservient to the Africans. Governing would be done by Africans, money would all be turned over to the Africans, and the authority would become that of the Africans. Indeed, the Africans would take over the nicer homes of the Belgians, traders, and missionaries and drive the occupants out. To our dismay as American missionaries, independence also gave birth to a lawlessness that knew no bounds. Militias, rebellious troops, and other vigilante groups were formed to "rule" certain regions and territories. At the same time, this spirit of *Uhuru*, Kiswahili for independence, took a wickedly dangerous turn. Many of the combatants felt they deserved to have the women of the Belgians, missionaries, and other ex-patriots working in the Congo, as well. Numerous atrocities were performed in the name of "independence," including theft, burning property, rape, even killings. Many well-intentioned, good-minded people, men and women alike, sacrificed their lives as the struggles played out. Quickly the Belgians turned over the government to the Africans, and many, many people fled the Congo in great haste.

The sudden and massive exit of the Belgians and ex-patriots left a huge vacuum in the government of the country. It is said that on the day the Belgian Congo received its independence in 1960 there were only eight Africans in the entire country who had attended school beyond the eighth grade. (Due to the enormity of the country, and its very diverse regions and tribes, it is impossible to verify this number.) Suffice it to say that the Congo was woefully unprepared for self-government.

Another issue was the question of who owned the mines, minerals, and timber in the Congo. Diamonds, gold, silver, rubber, uranium, cobalt, and other precious metals were being mined in the Congo and shipped around the world. Meanwhile the African mine workers were paid a pittance for their difficult work, done often by hand in sweltering tropical heat. Since these industries were owned mainly by the Belgians, who owned the wealth after independence? Numerous foreign countries moved quickly into the vacuum, while at the same time influencing the new Congo government by installing a leader, prime minister, president, or governor sympathetic to the interests of the foreign governments. From the West to Europe to the East came outside influences, mainly with their own interests at heart. With so few educated Africans, government and power became a free-for-all. Assassinations, exiles, and rigged elections became commonplace. The "new" Congo really didn't stand a chance politically.

Into this maelstrom of cascading events were thrown the missionaries, schools, and families of the Berean mission working in the jungles of the Congo. Many of our friends and fellow workers from other missions and faiths were also set adrift in the escalating trouble. For our part at mission Katanti, and the other four Berean mission sites in the Congo, things changed rapidly. While the African Christians defended us and took a stand for the missionaries, at any time and any place we could be harassed and attacked by African militant, rebellious groups. Feelings and emotions ran high, not just within the American missionary communities. For their part the Africans had many impossible choices to make themselves. The mantra of the day and time was to always be watching and listening, never letting down one's guard.

Someone in America had given the missionaries a network of five shortwave transmitter/receivers, so that at a given daily time, voice communication would be made between the five mission posts. In that way, keeping in daily touch was a source of encouragement to those of us still living and working in the bush country of the Congo.

We missionary kids had been raised for most of the past ten years in the Belgium-owned country of Congo. For no other reason than that our skin was white, we had been looked up to as friends, mentors, teachers—guarded, cared for, and yes, obeyed. Suddenly, the winds of independence, *Uhuru*, changed all of that. Many of us mission kids wavered in our feelings for our African friends, a wavering that would become downright hostile in the days and months to come. As these emotional and political waves swept the country, it became a more and more dangerous place to stay.

Our missionary parents finally made a decision that all white missionaries would evacuate by a caravan of cars, down the infamous "Route de Kimbili", through which we had passed many times, to the neighboring border crossing into Rwanda at the Congolese town of Bukavu. A day's drive northeast from Bukavu was a Rwandan missionary retreat center known as Kumbya. It would provide a safe place for us to go while we waited to see what would happen with the political climate of the Congo. In the interest of safety, we sometimes traveled at night, hoping to remain "invisible." Our efforts were futile, as you will see from the story of our journey to evacuate.

CHAPTER FIFTEEN
TROUBLE AT OUR DOORSTEP(S)!

From Dad's book:

"'GET UP, DAD. They're here!' It was David's voice calling into the bedroom where we had just stretched out for our afternoon siesta, an African custom. For weeks and months we had been harassed many times by marauding soldiers and other militants, and had always been released without harm. One could tell this time was different, more dangerous and urgent. 'Who's here?' I replied. 'The soldiers, they are all over the place!'

The soldiers had roared onto our mission in three trucks, armed to the teeth with loaded guns, and had spilled out of the three trucks. As was their custom, their first item of business was to jump down and urinate all over the mission property. Jumping up I told all the girls and mission ladies to get upstairs and stay out of sight. Going to the door, we could see Dave was right. Our homes on the mission were surrounded by gun-wielding, drunken soldiers. These were from the Congolese army, basically under the command of no one. The soldiers stood in front of each house, their guns cocked and ready. One of them came up to the door to speak to me. 'We were told you have a gun here, and we have come to confiscate it,' said the wild-eyed soldier. 'You know all guns are to be turned over to the government.' 'We did that,' I replied. 'We have no guns here.' 'Oh, yes you do!' exploded the soldier, taking a step nearer. 'We have been told one of the missionary boys has a gun he uses to hunt in the forest!' My heart sank! That pellet gun one of the boys owned. An African had turned us in for having what he thought was a real gun. Trying to laugh I said, 'That's only a toy gun.' 'Oh yeah,' the soldier roared in anger, 'we want to see that toy gun.' The soldier followed me to where the pellet gun was in a neighboring house. Suddenly he dropped to one knee and took aim right at me. Seeing that, an African pastor stepped between me and the rifle aimed at me. Reaching the porch, I called out, 'They want the pellet gun.' Gruffly, another soldier started to push past me to go into the house where the pellet gun was. Instinctively, I barred his way with my arm. He became very angry and another soldier at the side of the porch aimed his gun at me, with his finger on the trigger. I noticed he was one who had been at our mission school a year or two before. 'I thought you were my friend,' I said quietly. 'I am!' he exclaimed, with his finger still on the trigger. 'Well,' I said, 'this is the first time I have ever seen one friend aiming a gun at another friend.' Sheepishly, he grinned and put down his gun. With that, the pellet gun arrived. Examining it, the soldiers agreed it was only a 'toy gun.' Piling back into their trucks they roared off into the jungle.

Slowly, one of the African pastors came over to us, saying, 'We saw the soldiers with their guns aimed at you. Maybe you didn't know that there were other soldiers with their guns trained on all of the other missionary men. They were stationed behind cars and trucks, at the ready. We pastors thought, are we going to have to stand and watch them kill our missionaries? As we did for you, so we did for the other men missionaries. We placed ourselves between their guns and the missionaries. The soldiers told us to move, but we refused. *Now we understand that you must leave us!*' What an act of courage."

Our caravan reached Mushweshe, a mission station about an hour out of the border-crossing city of Bukavu. After driving all night, we were bushed. The people there made up some makeshift beds, and it was decided we could sleep a few hours, before heading for the border crossing. As we awoke and prepared to leave, a different group of rebellious soldiers arrived to shake us down and harass us again. This time they wanted to go through all of our suitcases. Even though we had left with just one bag each, this took a long time to accomplish. As for me, I suddenly had a knot in my throat, making it hard to breathe. I knew the soldiers would frown upon any weapons we had in our belongings. I had hidden my long-time favorite razor-sharp machete in between some clothes in my bag. As they opened my bag, I could not look my watching parents in the eye. Holding my breath, I watched as the soldier looking through my bag went right on, without saying a word about my machete. I could not look at my parents, who were looking daggers at me. Lesson learned.

Finally, we headed down the mountainside toward the border. In the town of Bukavu, again we were stopped by truckloads of soldiers. This time they wanted a bribe to escort us safely to the border. We obliged, and arrived across the border into Rwanda safely.

At a missionary retreat center, some of those with us stayed in cabins, while our family stayed in a stick-and-thatched hut. The retreat was built on a peninsula into the beautiful Lake Kivu. Frankly, we were happy to have safe shelter. While we were there for two or three weeks, word would come by shortwave radio about the situation in the Congo. Finally, Dad and two other men made a trip back into the Congo mission region to see for themselves. Since the conditions were somewhat improved, each missionary family had a tough decision to make: either go home to the United States or return to the jungle stations in the midst of a fractured, dangerous country. Our family chose to go back into the Congo. Some of us kids were not too happy and went back grudgingly.

We weren't back at our mission too long before another call for evacuation became necessary. Some American missionaries had been brutally raped, beaten, and massacred near Stanleyville, a town in our region, just one day's drive from us. In fact, the American consulate was broadcasting over shortwave for all Americans to evacuate the Congo.

CHAPTER SIXTEEN
FLEEING AGAIN

ONCE AGAIN, WE said goodbye to our friends at mission station Katanti, in the Kivu rainforest. Again, we started the long, rough drive over the Route de Kimbili, to the border town of Bukavu. Someone had seen truckloads of soldiers heading in the same direction as we were. We had no idea if we'd meet them or be ambushed along the way. We decided to drive at night, hoping to cloak our escape in darkness. At one stop along the mountain road, someone thought they heard sounds up on the mountain above us. Hastily, we drove on into the night. All of us were scared to death.

Coming around a corner we saw car lights that appeared to be coming toward us. Dad, who was beside himself and panicked with fear for his family, shouted, "There they come." He was so concerned for us and our entire party that seeing those lights could mean nothing but soldiers. No one spoke, afraid to break the tension and alert the soldiers coming. Suddenly someone realized the lights were actually going in the same direction we were; a hairpin curve had made it seem that they were coming our way. Dad was so tense and exhausted he had to have some relief. Mother gave him some aspirin, and he rested leaning against her. David drove the car as we continued onward.

Arriving at mission Mushweshe, where we had rested on our first evacuation, we agreed it had never looked so good to us. We tried to rest until morning; however this post was very close to the hostilities, and rest wouldn't come. In fact, the missionaries at this post were themselves making plans to evacuate.

From Mother's book:

"The next morning the men went down the road toward Bukavu. Their goal was the United Nations post in Bukavu, the border city. The UN requested the men bring all of our party to the United Nations compound in town, and stay with them while they tried to arrange our passage across the Ruzizi River crossing into Rwanda and safety.

Starting down the seven miles of mountain road we came to the first road block. A tough sergeant ordered us out of the cars, insisting we open every suitcase. Satisfied that we were not hiding weapons nor money, he asked for a bribe, we delivered, and he waved us on our way. At the next road block they asked to see our passports, along with another bribe. When we finally reached the peninsula in town where the United Nations was headquartered in several houses, they welcomed us with handshakes and smiles. God had provided UN troops from Nigeria who spoke English.

In the morning the UN commandant brought us news. He and the local official from the U.S. Consulate had gone to see the local African regional governor, a rebel himself, named Kashamura. They carried a request for our mission party to be allowed to leave the Congo. As Kashamura was a known troublemaker, rebel, and rabble rouser himself, he demanded that we write out written requests for each person. Further, he demanded the requests be written in French, signed and returned to him for his signature. We lost no time doing his demands, but in short order the messenger who had taken the stack of requests to Kashamura was back.

The governor had refused to sign them, saying that we had to rewrite them all, stating that we definitely had no plans to return to the Congo. This put a different light on things for us missionaries. Many had hopes of returning to our stations after the trouble was over. My husband was among that number. To sign a statement that he had no desire to return could not be done lightly. After prayer and discussion it was decided to go ahead and rewrite and sign the requests. We reasoned that if Kashamura and his rebellious soldiers were successful, we would not be allowed to return, anyway. If they were not successful, these papers would be useless."

My memories of this evacuation are still very clear, so I will pick up the story myself. On the second day, around noon, the UN commandant came running. He had good news, or so we thought. He told us we were going to head out to the border crossing, with the requests for our clearance in the hands of the UN. We were to be at the border crossing in two hours. A UN officer went in a Jeep at the head of the caravan, and our car was the last one in line. Many, many people lined the streets of Bukavu, yelling, shouting, and waving at us. It was hard to know friend from foe in those surroundings.

By the time we arrived at the Ruzizi River crossing there was already a loud argument between Congolese rebel soldiers and the UN people. To our dismay, there was much shouting and waving of guns. The weapons were carried by Congolese soldiers, most of them drunk and unstable! The rebel soldiers would not even look at the documents signed by the governor and held in the hands of the UN people. Instead, all of us missionaries, about twenty-eight including the kids, were herded roughly into a round metal building. It was hot from the tropical sun, and we barely had room to breathe. Now we were in the custody of drunken African soldiers, brandishing obviously foreign-made weapons.

The argument for our fate continued for a long time. While it continued, we breathed silent prayers for our safety. Suddenly the UN commander turned to his Jeep driver and told him to turn his Jeep around. The driver refused, fearing that the rebels would shoot him. The commander repeated his request, and jumped into the right-hand seat. The driver whipped the Jeep around and roared off in the direction of town. Someone in our group of captives whispered, "Oh, no, the UN is leaving us here alone." No one knew what to do next.

In a few minutes more trucks showed up with more angry rebel soldiers. They surrounded our tiny metal shack where we were being held. These rebels looked wild-eyed and drunk, waving their weapons and shouting at us as if we were a group of vicious offenders. Just then the UN Jeep with the commander roared back to us. Again, a Congolese and a UN man tried to reason with the rebel soldiers. The effort was futile. As the UN man walked up to the command shack, a soldier strode to him and pointed a loaded gun right at his chest. Shaken, we thought we were going to see a shooting right before our eyes. The UN man stopped, staring right into the eyes of the rebel. Nothing happened. After a few seconds, the UN man took another step toward the rifle pointed at his chest. Again, nothing happened. Taking a third step, he reached out and pushed the gun to one side, telling the rebel soldier, "Put that gun down and let's talk this over." For just a few moments we breathed a sigh of thankfulness and relief.

Without warning, we were roughly ordered out of our metal hut. Women and girls were to get into a van driven by the rebels. Missionary men would follow in our own cars, wherever the rebels were taking us. These instructions were very scary for us missionary captives. Many were the stories we had heard of missionary women and girls being brutally raped and kidnapped or killed by these rebels. Others had been forced to watch their husbands and sons killed, before being attacked themselves. None of us felt at all comfortable, as fear rose in our hearts and souls. As I watched Mother and sister Patty and other women and girls being separated from the men, my own ragged breathing caught in my throat.

Among the pandemonium of the moment came a new order: "Every family get into your own cars, with an African rebel soldier with you, then, follow the lead vehicle." Breathing a sigh of relief, we complied, wondering where we would go from here.

The line of cars and military vehicles drove to the Hotel Riviera back in downtown Bukavu, Congo. The Riviera had been a plush hotel right on the shore of Lake Kivu, a favorite of the Belgians in their day. It now was the headquarters of the fractured local government, and Kashamura. All twenty-eight of us were roughly manhandled into two lines in front of the hotel. Women and girls were lined up on one side and men and boys on the other. With fear and trembling, we did exactly as we were told to do. With governor Kashamura nowhere in sight, a rebel soldier yelled out, "What language do you speak?" Dad answered that we spoke French and Kilega, one of the local dialects. A soldier whose face bore the markings of the Balega tribe asked if we could really speak Kilega. Sensing a ray of hope, Dad exaggerated. "Yes, we speak all the Kilega there is to know." To our relief, the soldier turned to his rebel commander, saying that we were missionaries and not mixed up in politics. Again, we breathed a prayer of thanks.

After a long wait, now on our third day of captivity, the UN commander returned waving the "permits." He announced we would go again to the border crossing, this time with a police escort. Dad asked why we had been arrested. The UN officer shrugged and let out a long, deep breath. Kashamura now demanded all the written requests be hand-signed individually! We realized that only the top one was signed. Each of us meticulously signed our respective requests, even the children in our group. The whole scenario was no more than a ruse, and harassment of the missionaries. Everyone knew that the rebels at the border had refused to even have a look at our so-called permits. We headed in our hapless caravan back to the border crossing.

The border guard removed the barrier and we headed across the rickety wooden bridge over the Ruzizi and into Rwanda. As the last car, which was ours, passed the guard on the way down to the bridge, the rebel held out his hand and asked for a "gift." I must admit, at age fifteen I was ready to give him a gift all right, just not the gift he wanted.

Just as we crossed the bridge into Rwanda, our hearts sank yet again. We saw that the fiasco had nearly started an international incident between Rwanda and the Congo. There, on the Rwanda hills overlooking the border crossing into the Congo, were hundreds of Rwandese soldiers, dug into the hills all along the riverbanks. Rifles and automatic weapons were trained on our caravan of cars. Cowering in fear, we continued to slowly drive up the hill into the relative freedom of Rwanda. Rwandan commanders told us their plan was to open fire on the border if the Congolese were to begin firing or committing atrocities. Wow! Our plight also had been flashed around the world and to big cities in America. Our loved ones in America had heard our names being read in news stories, including the fact of our capture by Congolese forces. In fact, it was weeks and months before some of our relatives knew we were out of immediate danger.

Arriving at the Kumbya retreat center where we had evacuated to a few months earlier, we were tired, scared, worn out, and just a little "crazy." For a number of days we twenty-eight missionaries talked, prayed, and debated our next move. Many wanted to go home to America. Others, like Dad, wanted to wait and see what would transpire in the Congo. The debate and talking went on and on, with no easy answers. Our mission group had many high-school-aged kids like my brothers and me. Schooling was an issue. Bad news from the Congo also was a big issue. While the missionaries discussed things, we kids had already made up our minds.

One night David and I requested to talk about the situation with our parents. This was what we said: "You know we have loved living in the Congo, but things are so different now. Many of the missionary kids are gone to America. There are only three or four of us left. If we did go back to Congo, it wouldn't be like it was before. Everything has changed. There is some talk of a man starting a mission school here at Kumbya, Rwanda. That is for the birds. We don't have any books or materials here. What kind of a school would that be?" We finished our talk with, "We've heard talk of all of us going to America. Really, that would be our choice."

After bedtime, Dad and Mother walked and talked deep into the night.

There were no easy answers. To missionaries who felt called to God's service in the Congo, it was difficult to make a clear decision. As for myself, I was *very* ready to chuck the whole thing and return to America. To be honest I had had more than enough of the Africans.

And it was not only the white missionaries who were harassed and threatened by the Congolese rebels. After the departure of the missionaries, the African pastors had left their homes and were hiding out in a guest house on the peninsula in Bukavu. Suddenly, a truckload of rebel soldiers roared into the yard, shouting, yelling, and aiming their guns at the African pastors. They rounded everyone up, standing them outside.

"Line up," the rebels shouted, "you are going to die! We are going to shoot you with our weapons."

The pastors pleaded for an explanation, telling the soldiers that they were God's preachers and doing God's work. Sneering, the soldiers retorted that they didn't know this "God," nor anything about his work. With that, they took dead aim at all the pastors, their wives, and their children. In response, one of the pastors took his New Testament from his pocket, holding it high for all to see. He shouted, "I am not a person who is a troublemaker. I preach God's Word and His Gospel."

Hearing that, the soldiers slowly lowered their weapons. Demanding the pastors' state work permits and state tax books, they looked them over and handed them back to the group. Suddenly, the soldiers jumped into their trucks and roared away. Thanking God, the pastors and families returned to their house.

They were all sleeping that night when they were awakened by unmistakable sounds, *rat-tat*, *rat-tat*, *bang-bang*, and breaking glass. Eight of the rebels had returned and were firing their guns through the house, breaking windows and yelling. Confusion reigned!

One of the pastors grabbed his little girl and found a small open window. After he squeezed them both through the window, they hid in the tall grass outside. Alas, it was no use. They were found and rounded up again outside with all the others. The pastors and families lost all hope for their lives. Trembling as they faced armed rebel soldiers in a lineup, for the second time in one day, they gave up hope. One of the soldiers shouted, "If you can pay us 200,000 francs, you will live. If not, you die!" Hunting through all that the missionaries had left behind in their evacuation, they found enough money to pay the ransom for their lives. The soldiers were somewhat appeased, but before leaving they roughed up one of the pastors severely. The pastors had no choice but to flee themselves into neighboring Rwanda. Miraculously, they all got across the border into Rwanda, with the help of some Christians from across the border.

Meanwhile, as the missionaries were at the Kumbya retreat in Rwanda in limbo, another strange thing happened.

From Mother's book:

"*Putt . . . putt . . . putt*. What was that strange noise? Turning on a flashlight I saw it was 4 a.m. Listening intently, I heard the voice of one of the men evacuees saying, 'Get up quickly and dress, there is a boat on the lake heading our way. I heard them say, "There are the houses of the missionaries." They must be looking for us.'

Grabbing my robe I dashed out the door to the room where the kids were sleeping. As I went down the walkway, a spotlight from the boat picked me out of the night in a most threatening manner. I awakened David, Roger, and Patty, telling them to dress quickly. I told them there was a strange boat close by on the lake, and it may be soldiers looking for us. Patty came with me to my room, and the two boys went with the men to find out what we were to do.

Shivering with fright and chills Patty and I waited in the small bedroom, while the boat seemed to be coming closer. Realizing that my frantic silent prayers were not comforting Patty, I prayed aloud for our protection and comfort. Patty spoke a touching prayer as well. Together we recited the 23rd Psalm. Just as it seemed we would burst with fear and waiting, the boys burst into the room. 'Mom, Pat, it's all right,' they said. 'The boat is a coffee boat heading down the lake to unload at a market. They lost their way in the darkness of night.' Patty and I went back to bed together, but there was no more sleep that night.

When my husband returned from searching for news about the Congo, we realized a decision must be made. My husband walked, prayed, and cried all night before reaching a decision that as a family we should return to the United States. The safety of our family weighed heavily upon him, after seeing and hearing of the escalating fighting and dangerous atrocities.

At last his decision was made. He would take his family to the safety of America, as soon as possible."

PART ONE
KIVU YEARS IMAGES 1950-1960

FIRST MISSIONARIES LIVED IN TENTS-1938

Small hut for Evil Spirits

Bwame Witchdoctor dislikes missionaries

Green brothers and Dad on bridge

Green family driving into Congo-1950

CROSSING RIVER ON "BAC" JUNGLE FERRY

FIRST CONGO HOME, MUD-STICKS-THATCH

Station Katanti missionaries-1951

David and Roger in "Kipoi" carriage

Our family in Congo-1957

"The Legends: David, Eldon, and Roger Green

Dorm students at Berean Christian Academy, Congo-1950s

PART TWO

TRANSITIONS IN AMERICA 1960-2011

CHAPTER SEVENTEEN

SAFE AT LAST

WITH OUR FAMILY'S decision made, we began to make arrangements to fly home to America. First, we arranged for someone to sell our Ford station wagon in Rwanda. Next, we made arrangements to travel home. With such short notice, we were routed through Belgium this time, quite an irony. At the huge Belgian airport we discovered that we were going to be taking our first-ever flight on a jet-engine-powered airplane, via Pan American Airways. Wow. What excitement! The plane, the crew, the flight attendants, the seats, even the food were a tremendous treat to those of us who had only traveled previously on planes with propeller-driven engines. In those days people actually dressed up to fly. As we took off for New York's airport, one of us commented, "This thing takes off like a homesick angel!" Never had we experienced such a short trans-Atlantic flight.

In New York we landed in a snowstorm. The jet bridges were not working, so we disembarked down the forward and aft staircases, right onto the tarmac. The wind howled relentlessly across the airfield. Just as we began a long walk to the terminal another jet drove past us. As it turned into its gate, snow, ice, and sand were thrown all over us. Wow. What a welcome home. Believe me, we were just happy to be on safe, American soil. Phone calls were made to relatives across the U.S. to tell them we were home safely. Arrangements were made for us to journey across the U.S. to visit churches, friends, donors, and relatives. Joyful reunions were held all across the country, even as the news from the Congo continued to worsen. It was plain we had made the correct decisions.

After trips to many different parts of the States, we settled in an apartment in the small western Nebraska town of Atkinson. This was where Mother had grown up, and her parents had a farm nearby. David and I began high school at Atkinson High, and Patty went to an Atkinson grade school. The local news media came to do an interview with all of us, which was beamed around the country. We boys thought that was pretty cool! Eldon was now enrolled at John Brown University in Siloam Springs, Arkansas.

In 1961, our family moved to Central City, Nebraska, so David and I could attend Nebraska Christian High School. We lived there until we had both graduated from high school and moved on to our own lives in Omaha. Patty, still in school, stayed with our parents.

Sometime after I graduated from high school, our parents received word that it could be safe to return to the Congo as missionaries. Patty would go with them. They would open a bookstore, mission center, and reading center in the border city of Bukavu. This placement would have them right on the border—should trouble break out again. To be honest, with my memories of the trouble we'd had in the Congo at the hands of the Africans, I was happy to stay on American soil.

So we exchanged sad goodbyes and then our parents and Patty jetted off to Africa, by way of Belgium again, to take up residence in Bukavu.

During that phase of their service in the Congo, my parents had to evacuate with Patty yet again two more times, making a total of four times they fled for their lives from the Congo.

From Mother's book a last time:

"'Whewwwww,' my husband whistled, crossing the familiar Ruzizi River and at the bottom of the hill into Rwanda. 'We made it.' 'Look,' I shouted, 'that hill is covered with Rwandese soldiers with guns, and they are running toward us.' The cars ahead of ours were slowing down. The soldiers were indeed firing on us! To our amazement the American consul had stopped his Land Rover and jumped out (making a perfect target), waving his hands in a gesture of peace. At the same time the consul's wife was vigorously motioning us to continue driving quickly up the hill. Slowly the shooting stopped, and we made it to the top of the hill. One of our group was driving a Volkswagen 'Bug,' and he came running to our car. While driving through the hail of bullets, some had landed just short, in front of our cars. Others had landed just behind or whizzed right beside our windows. Continuing on to the airport at Kamembe, Rwanda, the U.S. consul ordered a C-130 to fly in to evacuate us out of Rwanda. Since there were no runway lights at that primitive airport, and, since no C-130 had ever landed there, we all lined up our cars along the airstrip with our headlights turned on, lighting the runway. Thank God our evacuation was successfully and safely accomplished!"

CHAPTER EIGHTEEN
LIFE IN THE UNITED STATES

FOR THE FIRST time, I was living in the U.S. alone. Eldon and David had married, and I was dating a wonderful girl from Illinois named Carol Williamson. On July 2, 1966, we were married in a beautiful church ceremony in Morton, Illinois. In 1968 our first child, Michelle, was born. When Mother, Dad, and Patty came home from Africa on their next furlough, they got to meet three daughters-in-law and three grandchildren.

Over the ensuing years I just forgot the Congo. Our family was growing, and "life" and its issues took my time. Our second daughter, Nicole, was born in Omaha, Nebraska. She had a birth defect called hydrocephalus. God sent us this special child, who lived with us for three and a half years. Praise God! In those three years our Nicki did everything the doctors had said she would never do. She was walking, talking, playing, and going to preschool, a joyful miracle of God. He called her Home on October 18, 1975.

In December of 1976 our family moved to Southern California. We now had added Danielle and Tony (Anthony) to our family. I landed a position as a manufacturer's sales rep with a national company. My job was a multi-line sales representative for over forty companies. We sold to Christian book and gift stores, and to secular book and gift stores too, with my territory the five western states. God granted us favor, and I rose to vice president and national sales manager. Africa and the Congolese were pretty far from my thoughts over the years. If I did hear something in the media, it seemed to me that the Congo in particular wasn't getting any better. I really had no desire to reopen that chapter of my life at all.

We lived and worked in Riverside, California, for thirty-plus years. Those were good years of growth and maturing as a family. I was privileged to help create a startup company with some dear friends, expanding what I had been doing. I will always have a soft spot in my heart for California.

In 2007 we sold our California home, moving to Dallas, Texas—specifically Keller, Texas. My wife, Carol, was experiencing neurological problems, which were a great concern to our family. We moved for Danielle and her family to assist with Carol's illness. In December 2007, Carol was diagnosed with an unusual type of dementia, Corticobasal Ganglionic Degeneration. The disease had indeed progressed, and the outlook was not good. Michelle and her family came to Texas to help out, and Tony and his family did all they could do, as well. I officially retired in 2009 and became Carol's caregiver in our home. Our church, Cross Timbers Community Church, Keller, and Pastor Brad Maughan were invaluable assets to us, as well.

Fast-forward to Spring 2011. The company I had helped to start, Integrity Sales & Marketing, Inc., owned by Tom and Vickie Propst of South Carolina, and Brian Bossman of Georgia, graciously made it possible for me to continue our medical insurance coverage, and to work as I could from our home, while caring for Carol. God and His people are so good!

While Carol battled against seizures, strokes, and dementia, she was called home to Heaven on June 28, 2011, from our home in Keller. Words cannot express our feelings, and we await anxiously the certainty of the time when we too are called to Heaven and reunited as a family. As the plaque in my home states: *God is Good, All the Time!*

PART TWO
TRANSITIONS IN AMERICA IMAGES 1960–2011

Roger and Carol Green in Texas-2007

CAROL GREEN AT HOME-2007

CAROL GREEN WITH LILACS, CALIFORNIA-2006

Roger and California fruit trees at home-2006

Green family Christmas, Texas-2009

PART THREE

NEW LIFE GOALS AND UGANDA MISSIONS

2011–2012

CHAPTER NINETEEN
ALONE AFTER 45 YEARS

AFTER FORTY-FIVE YEARS of marriage, I found myself alone again. Through the months of grieving, I formulated a plan for my life moving forward. Who am I kidding saying I formulated a plan. God is the One who brought me to a plan for living alone.

After a few months, I began traveling around the U.S., seeing old friends and family. At the same time my church in Keller had a service featuring a Pastor Dongo, of Uganda. When I met Dongo personally, and we shared some time together, I felt a moving of the Spirit in my heart. Pastor Dongo invited me to visit his ministry, Buyamba, Inc., and God Cares Primary, High School and Orphanages, in the capital city of Kampala, Uganda. For the first time in years and years I had an interest again in Africa. I could hear Carol's voice, as well. On one of her good days toward the end of her life, she awoke to say these words to me: "Rog, when I'm gone, you go back to Africa! I really never wanted to go there, but you must go!" I took that as a direct sign from God.

Using my California contacts I was able to meet the United States director of Buyamba, Inc. and speak with that organization on numerous occasions. The groundwork was being laid for my quite unlikely return to Africa.

So it was that in February 2012, I drove to California and joined a team of eight short-term missionaries on a mission trip to Uganda. (I had visited Kampala, Uganda, several times in my early years, growing up in Africa.) The trip was set to last ten days, and included the U.S. administrator of the mission, as well as the U.S. chief financial officer. The name *Buyamba* in the Lugandan language means "to help." God Cares Uganda was capably operated by the late Pastor Dongo, his wife, and his family in the capital, Kampala. At that time over fifteen-hundred African orphans of AIDS, poverty, war, terrorism, plagues, genocide, and other atrocities were being cared for and were provided a safe haven in a Christian environment and school. Indeed, Pastor Dongo was from the same background as the children. He and his lovely wife, Florence, cared for sixteen kids in their own home, along with running the schools and orphanages. The Ugandan kids dubbed me "Rogers," to go with my other list of names.

During my first mission trip back to Uganda, we traveled to bush villages. At one stop our host, the late Pastor Dongo, said, "Stay close to me here. The father is militant Al-Queda. He does not like Americans and has no use for the Gospel. Most of the time he is gone. He only comes home to steal his wife's money, or for sex. However, his wife does attend the bush church. We are 'working on' this family."

"Wow", I said.

After getting out of the car, we were met by a machete-wielding, sullen looking African named Shafique. It appeared he was not impressed with us. As I tried to greet him with a handshake, he ignored me. Regardless, I felt drawn to Shafique.

After we greeted his mother and saw her small hut, Shafique suddenly appeared at my side. To my surprise he spoke to me.

"Come with me into the bush", he said—still waving his machete.

Not feeling comfortable going with him, I silently said a prayer for safety. "Lord," I prayed, "I have no idea what lies ahead. If this is an opportunity you've given me, I'm willing to go. Guide my steps and make me a blessing to Shafique. In Jesus' name, Amen."

I followed Shafique up a slight hill, into the bush. We were completely alone together, and out of sight of the others. After a reasonably short walk, we approached a small bush hut. Sitting inside was a young woman holding a baby. Shafique proudly showed me his first child! While I had feared the worst, he had decided to take me to his home.

Breathing a prayer of thanks, I asked God how I could bless this young man. On impulse I took Shafique's free hand in mine. (In Africa men holding hands is a strong sign of friendship and trust.) We walked back to the road, where he allowed me to take a picture with him.

As we invited him to the bush church, he let me pray for him, his family, his father, and his mother. As Pastor Dongo often said, "God is good all the time, and all the time God is good."

What a blessed experience.

This first trip in 2012 did everything to encourage me, and nothing to discourage me from returning to Congo after fifty-one years. My resentment of the Africans was gone, and I had a great time with the kids of God Cares.

Flying into Entebbe International Airport, Uganda, brought back memories of the hijacked plane that was taken to Entebbe airport in 1976. Among its passengers were an Israeli sports team. Israeli commandoes raided Entebbe airport to rescue their people. The story is immortalized in the movie, *Raid on Entebbe*. To this day, the old terminal building and some of those planes are standing on the tarmac next to the new buildings. Quite a sight.

Since Kampala and Entebbe, Uganda, put me about a two-day drive from the Congo, my "homeland," I began to pray and prepare for a trip to return to the Democratic Republic of Congo, as it is now called. The more I prayed and planned, the more excited I became. To be honest, God had put a new desire in my heart, and I could barely wait to get started.

PART THREE
UGANDA MISSIONS IMAGES 2011–2012

Roger and Pastor Dongo, God Cares Schools, Uganda-2012

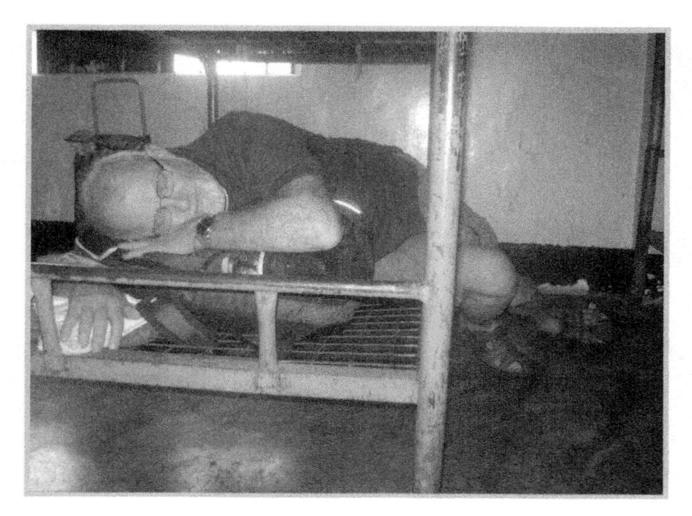

Roger—sleep wherever you can! Uganda-2012

Feeding 700 kids at God Cares, Uganda Bible Camp-2012

Watching "Futbol", (soccer), in Uganda-2012

Slum home by Kampala, Uganda sewage-2012

Hut in bush, drying grain near Ntunda, Uganda-2012

PART FOUR

UNLIKELY RETURN TO CONGO DRC

2012

CHAPTER TWENTY
A NEW VISION

DOING MY RESEARCH I found that two missionary couples, Tom and Kathy Lindquist and his brother, Jim, and Jim's wife, Louise, were involved in mission work in Bukavu, Congo DRC. This was a great break for me. I had grown up with these two guys in the dormitory in the Congo fifty-two years earlier. I had no idea they were still doing missions in the Congo at the time I became interested in returning to the Congo.

So it was that I contacted Tom and Kathy via phone and Skype to discuss the possibility of a return trip into the jungle. To my amazement, Tom said, "Rog, I will get you into the jungle. In fact, I will get you all the way 'home' to Jansen Hall at Katanti! None of this will be easy, in fact, it is dangerous in the Congo, but if you are really up for it and willing to bear the pain and struggles, I will get you there." It was music to my ears. After agreeing on the cost and logistics, we were making great progress.

My next scheduled trip to Uganda and God Cares was set up for August 2012. Providentially, Tom and Kathy had a trip of their own to the Congo DRC, for the end of August. The two trips would dovetail one behind the other in perfect fashion. My job was to travel alone from Uganda across Rwanda, and into the Congo DRC. Tom would meet me at the border to assist in immigration and border procedures. What an answer to prayer. Incredibly, my kids were also on board. All of them wished me Godspeed and gave their blessings. Was there danger? Yes! Uncertainty? Yes! Was there war and numerous rebel militias operating where I'd go? Yes! Was I willing to go? Yes!, Yes!, and Yes!

So it was that in August 2012 I flew from Dallas–Fort Worth to Entebbe International Airport, Uganda, with a stopover in Amsterdam and Kigali, Rwanda. I joined a team of short-term missionaries from Cross Timbers Community Church, Keller, Texas. Among that eighteen-member team was Bob Veach, my oldest daughter Michelle's husband. What a gift to do missions with him in Uganda at God Cares for ten days. The plan was for me to leave the team on my own at the end of our mission to Uganda and travel alone to the Congo DRC. I felt absolutely certain that Carol was joyfully watching me from Heaven.

Our time at God Cares Uganda went by quickly, with our team leading a Bible camp for seven hundred-fifty students. One of my former vendors from my sales career, Kerusso Activewear, donated eight hundred Christian T-shirts, rings, and jewelry for the African schoolkids. Two others, GT Luscombe and Universal Designs, also donated Christian pens and school kits. A fourth vendor, Gregg Gifts, Inc., donated Veggie Tales toys and shirts. What a joy to bring these things to needy kiddos in Uganda. My excitement grew daily as the time neared for me to go into Congo DRC.

CHAPTER TWENTY-ONE
AFRICA, MY NEW GOAL!

IT IS VERY necessary to have what they call a "protocol," i.e. a local person, to assist with everything in the Congo DRC. This includes visa, passport, something called a "Go Pass" for movement in country, a million "mystery" papers, and someone to accompany me on my travels for personal and safety reasons. I had not spoken Kilega, Kiswahili, nor French for fifty-two years. Communication would be an issue. Congo DRC, is on the U.S. State Department Watch List and No Travel List, which means that things are amazingly unpredictable day to day. One travels at one's own risk, permit or no permit. In country there are few if any means of communicating with the outside world. Fighting broke out just twenty-five miles from where I'd be crossing into the country a week before my arrival. At least eighteen different militias operate just in the territory where I was headed.

On Saturday night, August 18, 2012, I went to bed too excited to sleep. Still in Uganda, I would leave the next day. The goal: traveling alone across Uganda and Rwanda to the Congo DRC. The thought of returning to my boyhood home in the Congo far overshadowed any concerns or fears.

CHAPTER TWENTY-TWO
CONFLICT AND CONTRAST

AUGUST 19, 2012 was a red-letter day. I say I was "traveling alone," although I was sure of the presence of God with me. I awoke at 5:45 a.m. to prepare for the transition day from Kampala, Uganda to Congo DRC. First, I would go with the mission team to Kabalagala Pentecostal Church, where Dongo was the pastor. Then, I'd leave the group and travel alone toward my goal. Getting ready, I was amazed that we had hot water in the bathroom at the Olympia Hotel, a place we dubbed "Hollywood Hilton." This was an unusual blessing on my last morning. It was raining, making Kampala town a mess. Rain makes a mess of the red clay and open-flowing and overflowing raw, nasty sewers.

In spite of it all, the church was full for a wonderful service. I stayed through the worship segment of the second service, then Glory, Dongo's son, drove me to Entebbe, to the airport. By the way, all of Africa is very sensitive to picture-taking, especially at airports, police and government facilities, and banks. What pictures I took I managed to quickly shoot with my cell phone camera. I picked up a "spy pen" from an online website, however, it didn't really work well, and was pretty obvious too. One must be very careful. Arrests, fines, and even jailtime are definite possibilities.

The drive to Entebbe took about an hour, and we arrived about noon. African airports have a tight and strange security system. First, as travelers approach the airport grounds, everyone must get out of their transport and walk through a security scanner set up outside in the middle of nowhere. This is guarded by armed guard, and operates rain or shine. Then one must walk through a second, somewhat "better" security scanner to even enter the airport terminal itself. Finally, one must wait until approximately three hours before his or her flight to move into the area of the ticket counters. However, before you can approach the counters, you must clear another security scan. After check-in one waits in a different section until two hours before their flight, when it is possible to go through the last security gate into the boarding area. "Hurry up and wait" definitely applies to Africa.

Partway through all these security scans, I got a lunch of boiled chicken and fries, a Nile beverage, and settled down for the time to move toward the gate area. It was Flight 206 on Rwandair to Kigali, the capital of Rwanda. It was scheduled for 5:15 p.m., but remember, *This is Africa!* Schedules really don't mean much.

My international phone had gone dead, because I'd run out of prepaid minutes. Since this was a terrible time to be without international phone services, I tried over and over to top up with new minutes. Alas, after trying via text and cell calling, I had no success. The phone was "locked" since the prepaid minutes were gone. I had to figure out how to manage without an international phone. In addition, I dropped my tablet on the floor and totaled the screen.

So, although the airport had sometimes on and sometimes off wifi, my tablet was of no use, either.

Rwandair 206 came in an hour late, so what else was new? It was a small Dash 8 Embraer Turboprop, in pretty good shape. Of the thirty-seven seats, about thirty were full. Finally, our scheduled 5:15 p.m. flight left Entebbe airport at 6:10 p.m. The flight to Kigali, Rwanda, was about fifty-five minutes, and time passed quickly while I was downing my peanuts and passion fruit juice given out to passengers. I also met a Christian young man on mission from Germany, and a Youth With A Mission team from South Africa. We landed and cleared customs without incident. It was now dark in Central Africa. Having traveled through Kigali before, I was glad I knew the process. Since my phone and tablet were dead, I needed to meet Peter. My "protocol" Tom had messaged Peter to pick me up by taxi and take me to a guest house for the night. But wait, there was a man holding a sign reading "Roger Green." This man was an airport porter named Thomas. He had heard from Peter that he was running late, so Thomas would stay with me until the taxi arrived. In the meantime, I exchanged $100 U.S. for Rwandan francs, at a rate of 625/1. As Thomas waited with me outside the terminal, I was thankful Thomas and Peter both spoke good English. Rwanda's official language is French. Since it had been over fifty years since I had taken high school French, I sure was happy to have English-speaking guys.

Peter took me through "downtown Kigali," eager to show me the city. We drove down a wide avenue that wound past the government centers, President Paul Kagame's offices, the Ministry of Agriculture, Ministry of Defense, and the U.S. Embassy. At last, we turned about twenty-five yards off of the main road to arrive at a guest house, booked in advance for me by Tom Lindquist.

In Rwanda, as in much of Africa, the compounds surrounding buildings are enclosed in high, sturdy fences, with heavy locks and a guard armed with a Kalashnikov rifle. This is mainly to prevent theft. On the other hand, the guard was asleep. I was met by the lady who owned the guest house, who spoke only "little English." But gestures did the trick. The guard spoke *no* English at all. Now where was that high school French? Although I knew some Kiswahili, it is not used in Rwanda.

The owner showed me to a downstairs room in a mud-brick building. It had a ceiling fan (YES!), a double bed, and the power outlet actually worked! There were two community-shared bathrooms down the hall. All of this was just "adequate" by my estimation. The price, however, was perfect, only twenty-five dollars, including breakfast. My stomach felt a bit weird, so I took one of my trusty Cipro pills to ward off any bad crud. Peter was scheduled to pick me up the next morning at 5:00 a.m. for a ride to the airport and my short hop flight to Kamembe, Rwanda, just eight kilometers from the Congo DRC. It was time for "lights out" and some sleep.

Later, a group of six people from the U.S. came into the building and were quite loud in the sitting room, making it hard for anyone to sleep. Finally, they went to bed, thank God. I slept fitfully, waking up every hour, concerned I might miss my morning flight or Peter might be late with the car, you get the idea. Sometime in this life I need to learn to just relax and trust God.

CHAPTER TWENTY-THREE
ARRIVING "HOME"

MONDAY, AUGUST 20, 2012 was a HUGE day for me, the day I would meet my protocol, Tom Lindquist, and cross over into the Congo. I was eager to get up and get going from the Kigali, Rwanda, guest house. My alarm went off at 4:00 a.m., and I pulled myself together. After a quick wash—hot water, shocking!—I took my stuff outside to the courtyard to await my ride in the pitch-black African night. The guest house manager came out to ask me if I wanted breakfast. Thanking her, I declined. The main gate was chained and locked, and the guard was nowhere in sight. My heart sank, but *This is Africa*.

Peter showed up with my ride right on time and laid on his horn outside the compound gate, waking up various chickens, goats, and the guard. He appeared from the darkness of his sleeping post and let the car in to pick me up. We headed to the Kigali airport, about a fifteen minute drive.

The country of Rwanda is much more concerned with building infrastructure and upgrades than Uganda, or especially Congo

DRC. The major roads are wide and clean, and the country is proud of what they've accomplished. All of this just a few years after the devastating genocide.

Once again, just as the sun rose in the east, I had to walk through five security gates and clear customs, immigration, and passport control to reach my gate. Upon reaching my gate, I discovered there was no food and no toilet in the gate area. My gate lounge was packed, with long lines of people. And then, still another security check. With my international phone still down, I pressed on toward my goal of meeting up with Tom. Because we could not communicate, I had no idea if Tom would even be at the border to meet me. Going on faith, I carried on.

At long last, and right on Africa time—an hour late—we were ready to fly to Kamembe, Rwanda. There were only ten people on a plane with thirty-seven seats. Flying time was a quick thirty-five minutes. En route we were once again served passion fruit juice and peanuts. The plane made a very hard landing at a primitive airport in Kamembe, just eight kilometers from the Congo DRC. On approach into Kamembe I was treated to great views of Lake Kivu and several Congolese towns. I recognized places I frequented as a kid, over fifty-two years before. Such emotions and great memories came rushing over me. At the same time, there was concern for what I would find in the conflict-torn country of Congo DRC.

The landing strip was very rough blacktop, no runway lights, no tower, no building lights, no baggage claim, and just a few shacks made of tin. There was a primitive, older-than-dirt security scanner, which no one seemed to know how to operate. The airport was "guarded" by four Rwandese soldiers with Kalashnikov rifles.

They seemed quite indifferent to the whole place, but when I asked for a picture they refused.

The taxiway and parking were all red dirt. The original terminal building had been ransacked by sectarian violence, genocidaires, and soldiers' rebellions and wars. It still stood unused, closed up with old pieces of tin and rotting away. I was told repairs of that facility were not high on anyone's budget. There was an old American-style four-wheeler, probably purchased from China. I soon found out that the four-wheeler was used to carry baggage from the plane. With no baggage claim area, the bags were just dumped in a pile on the ground, rain or shine. Passengers had to dig through the pile to retrieve their bags personally. Next to the terminal shack was what might generously be called a rudimentary coffee shop. But no coffee. At least the toilets did work.

As I retrieved my luggage, a man named Desant got out of his car and introduced himself. He was a taxi driver Tom had hired to take me to the Rwanda/Congo DRC border. The short drive from a high plateau to the Cyangugu border crossing at the Ruzizi River took just a few minutes. My time was spent drinking in so many memories. At the same time, I was overwhelmed with lots of scary memories, as well. On this very spot and at this very border crossing was where our harrowing, dangerous escape had taken place fifty-two years ago! Our very lives had been in the balance. My vivid memories brought goosebumps the size of peas on my arms.

Arriving at the Ruzizi River, I found a shack on the Rwanda side for customs and immigration. I stood in line under the tropical sun, waiting my turn. Although my driver and the customs people knew no English, I was cleared through without any problems. During that time, my protocol, Tom Lindquist, arrived on the Congo DRC side to meet me. How good to see his smiling face. We picked up my backpack and one medium-sized bag and walked across the Ruzizi River on a broken-down bridge teeming with people on foot. After all those years, I was back in the Congo DRC. Amazing.

CHAPTER TWENTY-FOUR
REUNION IN CONGO

NOW I HAD goosebumps the size of eggs as Tom led me to the small, dingy immigration building on the Congo side of the border. Tom knew some of the people working there, so things went smoothly. I had to "appear" before a lady official for her to "approve" me into Congo DRC. Since Tom also knew Kiswahili, French, Kilega, and English, things went smoothly for me. After clearing customs on the Congo DRC side we walked up a hill and took a taxi to Tom's apartment in Bukavu, the border town. I had traversed this road many, many times before, and frankly, it looked to me like the place had regressed about fifty years from what it was in 1960. How sad.

We got out of the cab and climbed rickety old stairs to the third-floor apartment where Tom, Kathy, Renee (daughter), and Ella (granddaughter) lived when they were in the Congo DRC. Tom and Kathy and his brother, Jim, and Jim's wife, Louise, split their time between their homes in the U.S. and their apartments in Bukavu. They traveled back and forth at least three times per year, to work with Berean churches and a school for the deaf in Congo DRC. I really respected their commitment. This was not an easy place to work, and being the Congo, anything could happen at any time. Still, they carry on. The Bukavu apartment had a living room, dining space, bathroom, two bedrooms, hallway, and a tiny kitchen. The building was built by the Belgians in the 1950s. On the street side was a balcony overlooking one of the busiest and noisiest streets in Bukavu.

I renewed friendships with Tom and Kathy, who I had not seen in fifty-two years, and met Ella for the first time. Basodecki, their African housekeeper, cook, and jack of all trades, welcomed me, as well. All of us shared many, many memories as we caught up on each other's lives. Bukavu town has about three quarters of a million people. I say "about" because there is no such thing as a census count, ever. Most of those people live in a state of poverty.

Basodecki served us eggs, coffee, juice, and toast for breakfast while we enjoyed laughing and talking. Once again I had to be very careful about taking pictures, a real downer for me. I really wanted to take a "ton" of them. However, Tom and Kathy told me that often visitors had been arrested, detained for up to three days in jail with no food or water, and fined up to $500 U.S. for "not having a proper photo permit." Never mind that there is no such thing as a photo permit! This was just one of the many ways to extort money from visitors, especially white visitors. Tom preferred that I just keep the camera out of sight, having seen too much trouble from cameras.

Well, after breakfast, I sneaked two pictures of beautiful Lake Kivu, from the balcony. Because of the smoke there from a gazillion fires, it is always hazy, and Lake Kivu now emits a methane gas (!), which they use for energy purposes. As a kid I spent countless hours swimming in Lake Kivu.

A young African man named Songa, a grandson of someone I knew back in the day, came to greet me and practice his English. I tried to practice my Kilega with him. His English was much better than my Kilega. One thing I had counted upon was for my African languages to return quickly to my mind. After all, I was the kid the Africans had often told, "Roger, *Banuamazi*, you speak our language even better than we do." Sadly, at the ripe old age of sixty-seven my recall button wasn't working, a major personal disappointment. I did have a chance to enjoy giving Songa a letter from my brother David in the U.S. We had a good laugh at some of the things Dave had written in Kilega.

When I first arrived in Bukavu the power and water supplies were both not working. Tom had set up a power system connected to a group of car batteries in his apartment. He also had a solar panel system to keep the batteries charged. This did enable us to have lights and power. Amazing. We went online with Kathy's laptop so I could put more money into my international phone, making it operational again. Happy Roger again. I texted my kids in the U.S. that I was safely in Congo DRC. While all that was taking place, the power and lights came back on.

Lunch consisted of rice, greens, beef cooked in palm oil, and soup. Terrific. It tasted just as good as I remembered from fifty years earlier. Some Congolese came by to buy some items from Renee, who was closing a bakery she owned next door. I practiced languages more with Songa.

As I reflected upon what I was seeing, I began to understand the plight of the Congolese people.

Yikes! What a mess the Congo DRC has become. This country should be one of the richest countries in the world, not one of the poorest.

Yet it is perennially listed at the top of "Poorest Nations" list. Congo DRC holds the world's largest concentrations of gold, silver, diamonds, tin, cobalt,and coltan. Coltan is used in every cell phone and computer in the world. In addition this country is home to the second largest rainforest on the planet.

Ongoing conflicts, meddling and intrusions by Western powers, warlords, local militias, corrupt government officials from top to bottom, neighboring nations, and even the Chinese and Middle East lead to a thriving black market on all of these minerals. The country's wealth is pillaged, plundered, and outright stolen. Dishonesty, thievery, smuggling, cheating, and plain old greed are rampant across the country. The rich get richer and the poor are at poverty level or barely able to survive. The minerals are taken out of the country at black market prices, or smuggled "free" across porous border posts and into the hands of greedy governments and entrepreneurs worldwide.

Meanwhile, the native African miners get next to nothing, sometimes only one dollar per week, despite working in deplorable and dangerous conditions. Often they dig by hand, and always with someone watching their backs. All of this has been going on for years and years, and has escalated since the day in 1960 when the Congo DRC got its independence from the country of Belgium.

Along the way there has been constant fighting, genocides, and attempts to break up the country. Congolese heads of state have stolen literally billions of the country's riches, putting them away in foreign banks at the expense of their own Congolese people.

Volumes have been written about Congo DRC and its years of conflict. Rapes, killings, wars, regional conflicts, millions of dead or displaced Congolese. Homelessness, AIDS, poverty, and Ebola are reasons for these scourges that run rampant in the Congo. My intention is not to focus on those things, but to guide you through my own years growing up in the Colonial Belgian Congo, my fleeing the Congo after independence came in 1960, and my unlikely return to Congo DRC and my roots fifty-two years later, at age sixty-seven. Frankly, I had to see for myself, and it wasn't a pretty picture.

As I talked with Africans and missionaries alike in Bukavu, my heart was breaking for the many, many friends I have who somehow, through faith, true grit, and perseverance eke out a kind of subsistence living. The irony is this: In spite of appalling conditions and broken hearts, many of "my people" are beautiful beyond words. Their unwavering trust and faith in God, love for one another, and plain old grit are absolutely amazing.

CHAPTER TWENTY-FIVE
CATCHING UP

LATER IN THE day, Tom, Kathy, Renee, and I went to a large "slum" area of Bukavu called Pageco. In this area, a large company called Pageco had been run by the Belgians back in the days of colonization. We were invited to the Berean Center, where Berean's offices were located. This place was in the very heart of Bukavu, at a traffic circle where it seemed the entire population of the town had gathered. Such yelling, honking, pushing, shoving, and shouting. The local police, armed with automatic weapons, tried their best to maintain some kind of order. No one seemed to be paying much attention to the *gendarmes*. Police were also always on the lookout for vehicles carrying foreigners. Since the local police and national army had not been paid for months and months, they had become very good at making up ways to harass and extort money from foreigners—but even their own locals, as well.

My thoughts went back to the times we had made many mad dashes through this section of the city when fleeing for our lives. I must say it was a bit disquieting. It was good to know that I had entrusted all my bitter memories to God, and I could trust Him. Certainly it was still a dangerous situation, but that didn't seem to matter anymore.

At the mission center many African pastors welcomed me with a traditional African ceremony. At least eight men came to make personal speeches, recalling years of mission work with my parents over the years. It was a moving and heartwarming time. Tom interpreted for me, as these men didn't know English. After their speeches of welcome, I spoke through Tom about my joy at seeing them again.

As we wrapped up our meeting, several church leaders told of their needs, both financial and material, and asked for Americans and me to come and help them. This is not unusual at all in Africa. In fact, much is assumed by the Africans about the supposed "wealth" of the white men, so I needed to be careful and thoughtful about my responses to them. Thoughts from my thirty-plus years in Sales came to me: "Never over promise, always over deliver!" Tom led us in a prayer and then we returned to the apartment.

A young African immigration worker named Emile came by the apartment. He had heard we needed help getting permits in order to go downcountry to where we all used to live. So he offered to help us. Praise the Lord! Emile stayed until almost 8:00 p.m., so we waited for dinner. It was considered rude to eat in front of someone without asking them to eat with you. Frankly, we didn't have enough food to go around, let alone feed yet one more. We had cheese sandwiches and pineapple for dinner. The fresh pineapple was locally grown, and was to die for.

Rich MacDonald came by also. He was a missionary who had worked with my parents. We talked another couple of hours. Rich lives in Bukavu permanently with his wife, Kathy (formerly Kathy Baylor), and owns and operates Radio Okapi, a Christian station in Bukavu. I had grown up with Kathy many years ago in the Congo. Her parents were dear friends of our family. Once again it was a good time of memories and renewing relationships. Since Rich ran the radio station, he had much more to tell me about the situation in the Congo.

Later I pulled out a hide-a-bed in the living room, hung a mosquito net, and went to bed. On my first night back in the Congo after fifty-two years, I slept great!

CHAPTER TWENTY-SIX
NOSTALGIA IN BUKAVU

MY FIRST MORNING in Bukavu I awoke at 6:00 a.m. and went back to sleep until 7:00 a.m. Then, as my hosts awoke, I took the first of many, many spit baths, from a basin with a washcloth. Bukavu's water still wasn't working, and neither was the electricity. No surprises there. Breakfast consisted of eggs, toast, fresh mango, and cups of coffee.

This day there were lots of things going on for Renee (Tom's daughter) and Ella, her little girl. They were scheduled to return to the U.S. Basodecki, the houseboy, came to start his day at 9:00 a.m. He walked about five miles each way to work. After lunch Renee and Ella left for a six-hour bus ride from Congo DRC through Rwanda to Kigali, to catch their flight to the U.S. This was a trip with its own perils. After they left, I got Renee's room and bathroom.

Emile, the immigration man who was helping us with papers to allow movement inside the country, came by to pick up our passports, visas, my letter of invitation from the eighteen Berean churches to visit Congo DRC, and other papers. The invitation letter was required in order to get a Congo DRC visa. Emile would arrange transportation and acquire each of us a "Go Pass," required for movement inside the country. It was disconcerting, to say the least, to realize that Emile now had taken all our documents and papers with him. We could only trust God that the man would bring them back intact.

It looked like we could make our trip downcountry to the rain-forest on the coming Friday. We would catch a cargo flight, which would definitely be an adventure, and the agent actually promised us we would have seats. Unbelievable!

Later, Emile returned to say he had set up our passage from Bukavu downcountry, and that he would accompany us to Kavumu airport outside Bukavu on Friday morning. Emile was really working for us, trying his best to head off any "surprises" we might encounter on the trip into the jungle. It was nice of him to do so, and we guessed he probably wanted a gift. We gave him food for his work, which was not an unusual gift. Little did we know the bumps in the road ahead of us, despite Emile's best efforts.

I talked with Tom for a while regarding the Berean African churches and how best to help them. Consensus was that helping would be very difficult, due to the chaos and danger in Congo DRC. The African church elders really "wish for the good old days," wanting full-time white missionaries to come back to live with them and help them. Since that is not probable, they want someone to open an official U.S. Bureau (office) to represent their interests and present their needs and projects to American churches for funding. Given the fact that I was unable to help in that way, the matter was tabled. From my viewpoint I considered the matter to be finished. I should have known better. In the days to come, these requests would be brought to me again and again, with mounting pressure each and every time.

The day had dawned partly cloudy and warm, and again very smoky. There are hills all around Bukavu, and the streets were teeming night and day with people walking, walking, and walking hither and yon. Kathy was busy packing for our trip, so Tom and I hung out together.

After lunch, a lady named Suzanna and her friend came to see me. Suzanna was born at our mission station Katshungu Hospital, and her husband was a doctor there. She was the daughter of SaJacques, who worked for many years with Dad and Mother. Suzanna had moved to Bukavu from the interior in 2001, after lots and lots of trouble in her area, and running to the forest for hiding and safety. Although she had buried most of her stuff in the forest, the soldiers had followed her, dug it up, and stole her things. She had lost everything. What a story she had to tell. It was great meeting her and talking with her.

That same afternoon I helped Basodecki carry heavy furniture up the three steep flights of stairs, to store at Tom's place. We were fortunate that two younger men came along and helped us with the heaviest items. On one of the trips I accidentally stepped squarely into an open, full, stagnant sewer. Ugh and ugh. Welcome to Congo DRC. I had to completely clean and wash my feet, socks, and shoes. What a stench and ordeal.

Word came to us that our trip downcountry was being set up as planned. Later someone would bring by a scale to weigh every single item we would be taking, including ourselves. These planes were just barely airworthy, much less if you overloaded them. We paid Emile $300 U.S. each for our passage downcountry.

Just as I was getting ready for bed that night we heard a knock and "*Hodi*" (Hello) from the apartment doorway. Songa had come by to see me again. I got dressed again and talked with him for a little while, then excused myself to bed.

By the way, Tom, Kathy, and I had a great storytime at dinner, reliving our escapades as missionary kids downcountry all those years ago. What fun! Afterward, I went to bed in my own room, which had been vacated by Tom's daughter. As I fell asleep I had to wonder what surprises awaited me the next day. I would find out soon enough.

The next day I slept a bit later, until 7:30 a.m. The morning report was for only two crises that day. For Congo DRC that qualified as a "slow day."

Water Crisis again: While we had saved and reused a good amount of water in our "crisis containers," we could not go forever without fresh water, especially with three of us using and reusing what we had on hand. Tom gave us the new rules for water management and conservation: Re-use bath water, re-use sink water, only flush "solids and brown water," re-use teeth-brushing water, conserve drinking water, and conserve coffee-brewing water. This was when I learned a new mantra: "If it's brown flush it down, if it's yellow let it mellow!" Things were not yet critical, and we could buy bottled water at a store, but that was an expensive way to go. We would do that if the situation were to hit "critical" by Tom's standards. The word on the street was that the water would be off indefinitely, as someone was supposedly fixing something. At that point we had enough water for two days, so we would conserve and wait things out.

Electric Crisis again: For the second straight day, no electricity from the town. Tom's battery system was still good and strong, thank God. If things were to get "critical," Tom and I would have to put up the solar panel to charge the car batteries. So far, so good . . .

Evenings in Bukavu were cool, and the mosquitoes appeared at sundown, as well. Kathy had a small Bunsen-burner-type device with a smelly candle of some kind, which she would light and put under her chair in the evenings. Fire code? What fire code? And thank goodness for the mosquito nets at night.

Later word came that we had an "African" chance—meaning a big, fat maybe—we could get a truck of some kind downcountry to drive to station Katshungu. We would have loved that, but *This is Africa*, which meant the chances were really slim to none. As the day wore on, we couldn't contact any African who knew anything about a truck we might use. Like I said, *Fat chance*. And the road was reportedly impassable anyway. Probably just someone who saw a chance and had their hand out for money. Someone always had their hand out for money. If they couldn't think of a good reason, they would just make something up, and then the long, drawn-out negotiations would begin.

However, word did arrive that we would be sleeping in Jansen Hall at station Katanti. That being the house I had fled with my family in 1960, this was indeed great news. While that house was built with mud bricks and had a tin roof, it was crumbling and had been stripped by soldiers numerous times. Still, I couldn't hide my excitement to see it again.

Toward evening a man named Flavien came from an orphanage in Kaziba, outside Bukavu. They had sixty kids, twenty were infants. These were the children of the wars, AIDS, malaria, dysentery, HIV, and other diseases. So sad. Flavien spoke English and Swahili.

We discussed the Buyamba Uganda Orphanage I support and visit annually, and he invited Kathy and I to visit his place when we returned from downcountry. Suzanna also came to bring Bibles, songbooks, and Leadership Seminar lesson books for us to take downcountry for our Leadership seminar. We would definitely have quite a load.

At 4:00 p.m. it was time to go to the Pageco area of Bukavu again and meet with the elders once more. Tom, Kathy, Suzanna, and I took an old taxi. There were thousands of people milling about near the grounds of the Berean Center. We met Pastor Kitoto and twenty-five others, who said they had waited around for us in vain the previous day. (No good reason, we had always planned this meeting for this day.) For the next forty-five minutes, we were given a list again of what they needed: funding, sponsors, permanent missionaries, ideas, and help.

Once again, with Tom translating, I spoke to them that I was not equipped to meet these needs. But again, I would hear about this in the future. We left the area before dark, as it was not a good idea to be out after dark in Congo DRC. Danger lurked everywhere.

All of us were hot and tired upon returning to the apartment. Words could not express what I saw that day in Bukavu town, and Lord, have mercy. While there are three quarter of a million people there, it was a drop in the proverbial bucket to the population of the country as a whole.

After dinner we talked more about the situations in Congo DRC, politics, and crises. Also, we reviewed my day's events and my responses to the Africans. Valuable information, as I would soon learn.

That night I fell into bed, tired but happy. What a day, and what blessings. This day brought us one day closer to our trip downcountry.

CHAPTER TWENTY-SEVEN
THE BIG ONE

FOR ME, "THE big one" did not refer to the giant earthquake everyone fears in California—you know, the one where Phoenix becomes oceanfront property. I awoke to building excitement for our trip the very next day downcountry. Decisions, decisions: packing, weighing, and deciding what to take and what would have to stay behind. We were allowed one checked bag each, and the rest in a backpack. Including all the supplies for the Leadership Seminar, we were not to exceed one hundred-sixty kilos. Plans were coming together.

I took plenty (I thought) of batteries downcountry for my international phone, cell phone (mainly for music), and cameras. Pictures in the forest would be easier, although I still needed to be careful and ask permission. Sometimes I could get an African to take the picture, and that would help me out with that issue. I had spent over $100 U.S. for international phone calls and texts. It surely added up fast. So I added yet another $200 U.S. to my

SIM card global so we would be covered until we got back from downcountry.

The view in Bukavu that day was the best yet. It was clear enough to see the mountains where the gorillas are, Lake Kivu, and into Rwanda. We could also see the volcano near Goma town. I looked out over the town for a long while, as the smoke began to build again. The twin crises continued too. No water, no power again in Bukavu town.

Around noon, Kathy and I walked to an internet café and copy place down the street. Yet another eye-opener on the streets of Bukavu. They were hot, dirty, and there were people walking everywhere.

I needed to scan and email my tickets for my trip from the U.S. to our church treasurer. Tom and Kathy had fronted the money and needed my church to reimburse them. When we left the store, I forgot my original plane tickets in the copier! Ugh. Basodecki, the houseboy, went back right away, somewhat with fear and trembling. Praise God, the pages were still in the copier, a huge blessing considering this was Congo DRC. We were able to retrieve the originals. Isn't growing old and forgetful just grand?

Emile came to say everything was ready for our trip the next day. He stayed, and stayed, as was his pattern. After dinner I sent texts, emails, and Facebooked prayer requests for our safety and ministry in the jungle. I also told everyone, including my kids, that we'd be completely out of range of any type of communication for "at least" nine days.

I got to bed around 10:30 p.m., and slept fitfully. Tom and Kathy were still packing and repacking. My excitement level was through the roof. The very next day was THE BIG ONE!

CHAPTER TWENTY-EIGHT
EXTORTIONS AND BRIBES

ON THE BIG day, "fly day," I was up by 5:00 a.m. after a fitful night. Emile and Kandoto, the plane agent, were to arrive at 7:00 a.m. with a taxi. Imagine our surprise when for the first time ever in Africa, they showed up early at 6:15. Time means nothing in Africa, but usually that means people will be late, never early. The taxi waited while we quickly downed some breakfast with Emile. Kandoto had gone to work. Next, we loaded up and took the taxi to Kavumu, Bukavu's airport. It was about thirty kilometers, driving along Lake Kivu. We experienced all kinds of sights and sounds on the drive. I could not express the contrasts of this country. Squalor right below a hill where the Congolese president comes to retreat in a spiffy place above the lake. Several soldier roadblocks (legal and illegal) and checkpoints right beside the beautiful Lake Kivu. Open markets teeming with people, and rickety tables selling gasoline in plastic drink bottles. Also there were boats of every kind plying their trade across Lake Kivu to Goma, right beside wooden dugout canoes.

Along the way we passed the United Nations compound near the airport. There were some twenty thousand UN soldiers in Congo DRC, on a "peacekeeping" mission. The difficulty for them was that "peacekeeping" meant they could not use their weapons to keep things calm. What a deal!

Emile went with us to the airport, smoothing the way at the checkpoints, and to help us with our documents and papers. At the airport there were two more drive-through checkpoints, each more stringent than the one before, but okay.

When we climbed out of the taxi at the airport, the taxi driver spun the wheels of the old, rickety taxi on the gravel and sped off without paying his taxi's "airport tax" fee to the police and airport soldiers. That produced much yelling, and the police kicked the side of his moving car, then threw big rocks at his windows. Undaunted, he sped away. We were left to avoid the developing tension.

We went inside a small tin building and got a big surprise. The airport had added an actual waiting room, with actual chairs, and an actual working bathroom! *Say it ain't so, Mabel!* Tom and Kathy were stunned to see the new facility. The airport is quite primitive, but sees lots of traffic. There are UN planes, aircraft that belong to numerous NGOs (nongovernmental organizations), and Red Cross planes, not to mention the daily cargo flights into and out of the country. Also, there are many airplanes old and older in various states of disrepair, or just plain rotting hulks. The most interesting of the old planes is a single-engine Antonov, made by the Russians and which used to be the means of travel downcountry. However, that plane had been grounded by some "authority" due to its excessive number of crashes. The Africans dubbed that Antonov the "Titanic"! When a plane crashes in the jungle it is usually left right where it lies. It becomes stripped of anything valuable and lies there rotting.

Well, *this is Africa* and our flight time came and went without any sight of our plane. We would fly on a Polish-made Let M-410. It would be flown by European and African pilots, and was not like anything you've ever seen. Now converted for cargo use, the plane was a huge mess, and somewhat rickety. Everything inside the plane had been removed. All bulkheads, seats and seat belts, lighting, air blowers, bathrooms, the inside walls—all that remained was the empty shell of a plane.

We waited for hours for our plane to arrive. Numerous Africans also waited with us, not to mention a mountain of cargo stacked on the taxiway. Two "Ministers of Something" also arrived to wait for the same plane, and we met the man who used to be the governor of the South Kivu District, which includes Bukavu town. Our plane finally came, two hours late, having first decided to fly to Goma, across the lake. Why? Just because they can! As I said, there is no hurry in Africa.

First the cargo was loaded, then the people. The Africans climbed in through the cargo door and made the best of it. Sometimes the best of it meant climbing on top of the cargo and lying on top of the load, with maybe a foot of reclining and breathing space. Remember, we had been promised actual seats. The two African dignitaries took precedence, and took two of the three seats. Tom and I shared a seat with no belts and no safety anything, and Kathy offered to climb on the top of the cargo just behind the two "Ministers of Something." I immediately offered my half a seat to her, but she steadfastly refused. Absolutely no pictures of anything were allowed. The cockpit was separated from the rest of the plane only by an old, old, well-worn, dirty curtain. The two pilots climbed in over us to get to their chairs for flying the thing. The European pilot apologized to us, as he knew we had been promised seats. The pilot had not known the African ministers were coming, and there was nothing he could do.

The Let M-410 was old, old, and older. Frankly, I didn't think it was any better than the hulk called the "Titanic" plane. It had two loud turbo prop engines mounted up high on each wing, the wings being on top of the fuselage. There were no announcements, no intercom, and one was just happy to be on the darned thing.

Did I mention there were cracks in the body of the plane big enough to see through? This plane was a real workhorse, but looked barely airworthy. In fact, the international agency that rates planes and airports won't even give it a thumbs-up.

I also noticed the loud unrest among the Africans in the back on top of the cargo load. It seemed no one was happy, except for the "Ministers of Something."

We breathed a prayer and took off at noon, three hours late. Climbing above the first layer of broken clouds, we really couldn't see much through the window. It was yellow with age anyway. About fifty-five miles later we arrived in the Maniema provincial capital of Shabunda. The airstrip was just a dirt strip carved out of a clearing in the dense jungle. It was all visual flying. There were no buildings, just cargo waiting on the ground. But thank God for a nice, soft landing. Hundreds of people were there and were in a festive mood, dancing, singing, playing music, and they just kept coming toward the plane! No, the welcome was not meant for us. It was for the "Ministers of Something."

We were in the heart of the forest in Shabunda, Congo DRC, peeling ourselves like sardines out of the rickety old airplane. We noticed that we had parked right beside a second cargo plane, which had broken down there the day before. Word was that parts had been ordered for repairs. No one knew how long the parts would take to arrive.

I had been to Shabunda town many, many times as a kid, and I was eager to make a current comparison. The first exciting leg of my return trip to "my house" in Africa was accomplished. Little did we know what was ahead. Our patience would be tested under pressure, for sure!

CHAPTER TWENTY-NINE

NIGHT, DANGERS, AND DARKNESS

WE HAD JUST landed at Shabunda, a regional town where the unrest was worse than we'd left behind in Bukavu. On this day, it appeared that the emphasis was on the arriving "Ministers of Something." Immediately, an immigration man met us three missionaries and insisted we must follow him to a mud-and-stick shack for immigration procedures.

Although we had somewhat expected this, we had no idea what was to follow. We showed him our papers, which made absolutely no difference to him. We were told we would need to be "processed," and the man produced a form that looked suspiciously like the ones Emile had prepared for us in Bukavu precisely so we wouldn't need to deal with this here in Shabunda. The immigration man's words were punctuated by the appearance of two surly-looking soldiers carrying automatic rifles and insisting that we must fill out these papers and be processed again. The officials told us it would cost $300 U.S. for each of us, and if we would just fill them out we could be on our way. Things were definitely reminding me of 1960, when we had fled Africa. Now we were being hassled trying to get into the country. My guy, Tom, did not want to pay these "fees," since paying bribes and extortions perpetuates the graft and hassle in this country. I agreed with Tom, especially since we had not planned financially for this eventuality.

While we sat in the boiling, humid, smelly, tropical heat of that shack, they questioned our papers, visas, passports, residency, and anything else they could think of to question. These local officials questioned us just because they could. Things were becoming surreal, and that didn't even come close to describing our plight.

Tom negotiated and called Emile in Bukavu to explain our situation. Amazingly, the cell phone found a signal, but guess what? The local officials refused to even speak with Emile. Turned out the local officials didn't like Emile, and also didn't like being told what to do. Our negotiations continued, until our price to be "processed" was down to $5 U.S. each. We stood on principle while they stood on their demands.

Occasionally the armed soldiers would return and parade around to intimidate us. So who would blink first? Tom had made a commitment to take me to my house at Katanti, several hours away, and he wanted to fulfill his commitment to me.

We waited for over three hours, being treated to many shows of authority, and men with guns, just because they could! Everything is a big issue in Congo DRC, *talk, talk, talk, blah, blah, blah*, and it sure was a good thing Tom was doing the negotiations. I certainly would have lost my cool. Finally, two local Berean pastors came to bolster our case. After all that time, we were allowed to just "register" our presence, without paying any fees. Significantly, it was late afternoon, and the officials wanted to go home. It felt like a big victory to me, since the officials had blinked first.

Gratefully we walked two miles to the pastor's house and church. His lovely wife fed us rice, *sombe* (greens), and pineapple. We found out that the pastor had motorbikes, called *piki piki*, and we were told we could leave soon. We decided to head out toward the closest Berean mission station, Katshungu, to see the folks there and to retrieve stockpiled gasoline from that station. We were told that it would take us only one and a half hours to ride to Katshungu, so we still had plenty of daylight. Or so we thought.

It took at least an hour for the *piki piki* drivers to show up. Then, as is always the case, much *talk, talk, talk, blah, blah, blah* about the price for them to drive us. Even the smallest things were huge problems. After agreeing on a "deal," the guys needed to fill their bikes with gasoline—another couple of hours lost. We were charged $20 U.S. each, after the negotiations had started at $50 U.S. each. The gas in Shabunda was about $30 U.S. a gallon, out of a plastic drink bottle. Where did they get the gas? Don't ask.

Finally we left at 5:00 p.m., about an hour from darkness. I rode on the back of a *piki piki* driven by a surly African; Tom and Kathy drove and rode on the second *piki piki*; and Songa, from the hospital at Katshungu, loaded our considerable luggage onto his motorbike. We also sent some things along with a man on a bicycle, to be delivered to our seminar location at Katanti later in the week.

We headed into a setting sun in the rainforest. My *piki piki* driver immediately began loudly grumbling and complaining that "the *Muzungu*" (white man) wasn't riding on the back of his *piki piki* "correctly." Frankly, I didn't know there was a "correctly." I thought my job was to just hang on. So, as Rhett Butler once said to Scarlett O'Hara, "Frankly, my dear, I don't give a damn!"

In just a few kilometers we were stopped at a good-sized river, the Lualaba, where the bridge was gone. We had to walk the bikes across on a log, wondering what crocodiles and other critters were in the waters below. There was a new and unique meaning to the word "road" in Congo DRC. "Road" means a narrow, winding path cut through the jungle. Yes, there was a road here, back in 1960. No one kept it up, so the jungle took over again. The jungle, like the sea, always wins. Couple that with a torrential rainy season, and it became a treacherous path.

This path featured slippery, stinky, red mud pits of brackish water, big enough to swallow a Volkswagen Bug. Since there were no paths around these giant puddles, we just had to hang on as our *piki piki* drivers drove right through the puddles. The brackish water got splashed all over all of us. Our drivers wore long, rubber knee-high boots. We had no such luxury. The trick was to not fall off the bike into the river, mud pit, or whatever. Our bikes fell several times. My driver and I slid off a log crossing a river several times. At least the water was trying to flow at those spots (gallows humor). After picking ourselves up, getting the bike upright and getting started again, then taking a deep breath, we roared off again.

Another time I was attempting to help Kathy across a pole on foot when I lost my footing and went into the mud and water again. Later that night I discovered my domestic U.S. cell phone had fallen off my belt holster and "drowned" in the river. Gone forever were my contacts, emails, and one hundred-fifty plus tunes. The jungle won again.

We continued on, three determined *piki piki* riders, through saw grass slapping our arms and legs, under or over fallen trees across the path, and bouncing over huge rocks. Not to mention the drops along the path when climbing mountain terrain or meeting other *piki piki* and bicycles with no lights whatsoever or the women carrying huge loads on their backs. I was convinced there wasn't even one stretch of smooth path anywhere in the jungle.

My "insides," backside, and bones were rearranged time after time. I lost count of how many times we had to dismount and push the bikes through the muck. Remember, all of this happened in the dark of the African night.

As I mentioned, my *piki piki* driver was a surly young man. His favorite expression was a scowl. No doubt you've seen the type. He constantly complained that the "white man" he was carrying didn't know how to ride on the back of his bike. He suddenly stopped the caravan and ceremoniously began to berate me in Swahili about my "lack of skills." He then announced he "would not take me even one more kilometer!" Tom told him to "*haka makalele yako*," which means *stop your laughing at the white man*. Put another way, it can mean "shut up" or "get over it." For good measure, as my driver lit a cigarette, Tom said, "And, by the way, that cigarette will kill you." The drama was on in earnest. I was transferred to the cargo *piki*, driven by Songa, a nicer young man from one of the mission schools. That was a fine solution to me.

Finally, we arrived at a crossroad called Lugungu at full dark, with still a long way to go. Tom's headlight on his *piki piki* was not working. We had a meeting among ourselves right there at the fork in the path. Our choices: sleep along the road on the ground or press on through the night with the highest mountains still ahead of us. So we tied an LED flashlight to Tom's *piki piki* and decided to press on to our destination.

My new driver, Songa, made this contribution: "The worst is yet ahead of us." He wasn't kidding. Rough, slow, rocky, bumpy, going up and up in the pitch dark, too late now to change our minds.

A new menace was upon us, it started to lightning and thunder up ahead. Tom kept reminding me that he had said, "I told you this wouldn't be easy." Finally, we reached the top of the escarpment, and took a break at the top. It was so dark we could hardly see our hands in front of our faces. Thank God it wasn't yet raining. Someone had named this mountain "Telephone Hill." There was a path leading to the very summit of the mountain. If one stands just right under a certain tree, through the wonders of atmospheric changes, one can receive a weak cell phone signal. What craziness!

From the mountaintop we were just about seven kilometers from station Katshungu, down the "short" side of the mountain. We passed what was left of a hydroelectric plant that several missionaries had built across a rushing mountain river. Even in the dark we could hear the rushing water below. Alas, soldiers had come through several times and for no good reason trashed the electric plant, just because they could.

After all that hassle we arrived at station Katshungu at 8:00 p.m. Over three hours to cover the route we had been told would take one and a half hours. We were met with a celebration welcome, singing and praying by Pastor Masudi and his wonderful family. Tom and Kathy had taken a wrong turn in the night and showed up about twenty minutes later. Exhausted and smelly, we were fed a supper of rice, greens, and fresh pineapple.

This station had some deep groundwater wells built by some of the missionaries. Amazingly, they still work. This meant we could drink this water without having to boil it for thirty minutes to an hour. We drank deeply and gratefully. Masudi's wife gave us each a pan of warm water to do a "spit bath" in our rooms to get rid of the mud, muck, and stench. Wow, that was nice! (She would then wash our clothing the next day.)

I had one room in the mud-and-stick house, the family another room, and Tom and Kathy a third room. The fourth room was occupied by Masudi's grown daughter and her kids. The rooms were small, with dirt floors and a low platform for lying on, and a mosquito net. I did have a window, with no covering at all. The "toilet," a hole in the ground, was "down the path" in the trees. Pastor Masudi had built a makeshift wooden platform upon which to sit on the toilet. God bless that man.

We went to bed around 10:00 p.m. and dreamed of yesteryear in Congo DRC. The bed was hard, and it began to rain buckets. The huge thunderstorm had broken over our heads. Memories returned of raindrops on rooftops in Congo DRC; however, this would make the paths even more treacherous, and the path on our trip to my house would be impassable. The outside paths would be slick, and there would be no way to walk to the toilet in these conditions. So thank goodness for the low window ledge. I could use that for relief during the storm. What would tomorrow bring? I had no idea . . .

CHAPTER THIRTY
RAIN, RAIN, RAIN!

SO THIS WAS the "dry" season? Why did it rain for three days? Like Tom said, "Roger, the dry season is really a myth. Congo DRC really has a 'rainy' season and a 'rainy-er' season." We had lots of time on our hands at Katshungu, watching it rain. The jungle path toward my house at Katanti would now be a thrill a minute, so we had to wait.

An old woman named NyaYoanne came to visit us one morning. She was the daughter of a man who had started the Bible schools with Dad many, many years ago. Although she was now quite elderly, she carried on like a real "hoot" of a woman. We had great times with her as we talked and laughed.

Next came Songa, the man who worked at the hospital and had driven the *piki piki* for me after the other driver had refused to carry me any farther. He came to pick up a computer he had ordered through Tom. Later, he realized it was programmed in English. Tom had told him that up front and he had ordered it anyway. Songa wanted to know if Tom could take it back to Bukavu and reprogram it in French. "Yes," said Tom, "but it will take me time and you will have to pay for the French program." So there was a long discussion as always. Tom and Songa figured it out satisfactorily.

It had rained most of the night, and it kept right on raining. Many more Africans came to see us. There was much laughing, talking, and praying. That was how they showed their appreciation for our working so hard to come and visit them. The lady from the day before, NyaYoanne, brought me a live chicken as a gift, carried in an old tote bag she got from somewhere. We decided to have it for lunch just a couple of hours later.

For our part, we were busy discussing plans. Our original plan to leave for Katanti that day had been washed out many hours earlier. Tom decided to get his nephew Phillip's motorbike working, so we could rent one less motorbike for our trek. Also, we had to figure out how to measure gasoline, since in Congo DRC it is sold in plastic containers like drinking water bottles. Our task was to figure out seven liters for each bike to take us to our destination. As I said earlier, everything is a big issue in the Congo! Working on Phillip's bike, Tom took off the carburetor and found it full of sludge. So he traded it out with a carb from another hospital bike.

Midmorning it stopped raining, so Kathy and I asked Songa to take us on a walk around station Katshungu while Tom worked on the bike. Touring the station on foot we viewed the primitive hospital still in use, making do with whatever they had and using African doctors and staff.

The pharmacy and clinic buildings were no longer in use. All functions had been folded into the hospital building. Missionary Dr. Zemmer's house, built all those years ago, was falling down, with Africans living there. The former house of two white missionary ladies was gone, for the jungle always wins. We also came across several hulks of trucks and cars, now overgrown with jungle. Finally, we saw a house built for Tom and Kathy years ago, now crumbling to the ground.

On the station tour an unexpectedly scary thing happened. Suddenly Songa stopped and said quietly, "We cannot go on up that hill. I hear rebel soldiers up there who have taken it over." We were all too happy to go another way.

Other lowlights: The Bible school building was still in use, but in awful condition. Soldiers had come and stripped the library and desks several times. At one place we came upon a gas-engine-powered flour grinder, put up by some NGO (nongovernmental organization). Amazingly, it still worked. Pastor Masudi had hand-built a palm nut press, which worked on a hand-crank device. First it pressed out the palm oil, then the nuts were heated and ground for palm candles. Still later, the husks were sold for making soap. A very clever invention.

Finally we came across a creek we dubbed "Gold Creek." Yes, the water was gold in color from all the gold dust in the creek. There was mining upstream, where they were said to be taking out gold in bars the size of cement blocks.

Because of the rain and travel issues, all my clothes were being washed by Pastor Masudi's wife. So I wore Tom's clean extra clothing. Tom is a size forty-four and I'm a size thirty-four, so it was interesting, to say the least. I was happy to have something dry to wear and just dealt with the size issue.

Our last project for this day of waiting was my camera. Now that I had lost my cell phone and all its pictures in the river on our way through the jungle, I was even more upset when I dropped my backup digital camera on the ground—right on its lens. Of course part of it bent a little. Tom and I fiddled with it for an hour and got it straightened enough for it to work again. Yay!

With Phillip's motorbike still giving us fits, we decided to stay another day, working on the bike and hoping the rains would stop so we could continue our journey.

CHAPTER THIRTY-ONE
HASSLES AND OBSTACLES

OUR "SPARE" MOTORBIKE still wouldn't run properly—frustrating. We sent out word for other piki piki drivers we could make deals with to drive us to station Katanti. As always, we hoped for a reasonable price, and we'd use the free gasoline we had stockpiled at the hospital. We hoped maybe we could ride again on Sunday, or maybe Monday. It all depended on the rain.

After lunch I went to my bedroom for a nap. Tom met some guys from the village. When I awoke Tom was asleep on a chair, and the men were gone. While Tom napped, I sat by the *piki piki* being repaired, just to guard the parts and tools. One really must watch things in Congo, for stealing is a big problem. Even Pastor Masudi's own kids had stolen from their dad. Basically, this is a lawless culture. The so-called central government just ignores the eastern provinces because they are "too hard to govern."

At Katshungu the missionaries' tennis court was gone, the soccer field was overgrown, and any and all thoughts of cottage industries besides the grinder were gone. With this location being on a much more traveled path to the interior, the danger from soldiers, militias, and rebels was much higher. Added danger existed because of the presence of gold mines nearby. Pastor Masudi had been there forty years and had fled for his life forty-five times. Masudi had also been on the soldiers' "death list" for years. One time when the missionaries were still there, Masudi had sheltered some of them in the forest for a month. That was done at his own peril, and that of his family. God bless this good man.

Later that day some men came to have a big discussion about the Christian Education department in their school. Basically they were asking for help. The "government" wanted the churches to run the public schools, and in return to have one hour daily to teach Bible principles. Problem: There weren't any female teachers (which would be advantageous) to teach. Why? Because the more than thirty women who had come to the last training school were forced to watch soldiers burn all the school materials in the schoolyard. Their village was pillaged and burned, as well. And that was the least of what could happen to them.

Midafternoon two soldiers came to Pastor's house for another "shakedown". One was nice, the other was drunk. The drunk one took Tom to task for "not telling him we were coming and getting their permission." The fact that we had our papers with us in appropriate fashion meant nothing to these guys. They said it was "against regulations" so they could extort some money from us. *Blah, blah, blah, talk, talk, talk* until Tom said, "We're sorry," (for nothing) and left it at that. The soldiers got tired and left.

A local radio station was saying we were coming to Katanti soon for a Leadership seminar. Apparently many, many people were already on their way, with some walking up to one hundred-fifty miles.

At the end of this long day it was still raining, so we stayed yet another day.

CHAPTER THIRTY-TWO
CHURCH IN THE JUNGLE

AFTER SEVERAL DAYS it stopped raining. On that day, breakfast was meager, as we were running low on the provisions we had brought along on the trip. We had sent more supplies on to station Katanti via bicycle. So we had coffee, cheese, and a little bit of coffee cake. A man named Ezra came by to make a nice speech to thank our parents and us for coming to bring the Gospel to the Congo DRC.

Since it had stopped raining, the pastors decided to ring the bell for church. It was later than the usual time, but more than one hundred-fifty people came to the service. Each of us *Bazungu* (white people) were "presented" to the crowd, and then we each said a few words of greeting. Great fun. A little of my Kilega dialect was slowly returning to my brain, so I was able to communicate a few African proverbs to the people, in their language. The congregation loved it.

The church service consisted of three songs, and their custom is to sing all the verses of every song. Announcements followed the singing, and a special song was sung by a local African. Pastor spoke on the "woman at the well." A closing song was sung, and then Tom closed the meeting with a prayer.

One big downer! While we were at church someone broke into the pastor's house and stole a bunch of his belongings. One reason for the theft problem is the prices. A tiny bit of sugar cost $10 U.S.; a tiny bit of salt was $5 U.S.; gasoline was between $25 and $30 U.S. a gallon.

After church in a school classroom, we visited the burned-out and trashed-out church, wrecked in the last soldiers' uprising. Next to the burned church building, we found Tom's mother's grave on a grassy hillside. She gave her life in service in Congo DRC. Finally, we visited the hospital and met a man from station Katanti who knew my parents. He also remembered that he had played softball with me as a kid in the 1950s. The hospital is run via solar panels hooked up to batteries and an inverter.

One special thing about the burned-out church. Many, many years ago, one of the missionaries had painted a huge mural of the "narrow road that leads unto life, and the wide road that leads to destruction" on the front wall inside the church. It was a tremendous painting. It is still there, despite years of wear and tear, and more significantly despite numerous burnings, pillaging, and plundering by the soldiers. What a testimony that they were afraid to touch the Christian mural.

I asked about the absence of monkeys and snakes. When I had lived there years ago, both of those were rampant and we often came across them. Now the monkeys have been hunted so much for food that they are at least three hours' walk into the jungle, for their own safety. Snakes are not as plentiful as they used to be either, and yes, people do eat some snake meat.

Following lunch, Tom worked again on Phillip's *piki piki* and got it running. Since it was already afternoon and it was a long ride to our next destination, station Katanti and my house, we decided to stay one more night and leave early the next morning. We also made a deal with some local guys to transport us via *piki piki* for $40 U.S. and eight liters of gasoline each. Wonderful! We planned to leave at 9:00 a.m. the next morning. Things were finally coming together.

CHAPTER THIRTY-THREE
MIRACLES DO HAPPEN!

THIS WAS THE "day of days," when we headed out again on our way down the paths to my African home. My driver was a young man of twenty-four named Ramazani. He was a friendly, helpful guy, and a great piki piki driver. He could barrel across a narrow, slippery pole over a river, with me hanging on behind him. Without any trouble we would be at station Katanti in about three or four hours. Well, think again, This is Africa!

We got up at 6:30 a.m., with people already coming to say goodbye. We ate a little something about 7:15 a.m. During that time I was seated in a wicker chair in the pastor's house. A poisonous scorpion came out of the chair, ran down my bare arm, and jumped onto the floor. Thank the Lord the scorpion did not bite me.

Pastor Masudi said a prayer, and we started out. The weather had dawned clear and sunny. Within two kilometers we stopped to redistribute weight and reset the backpacks on the *piki pikis*. Then Tom's bike would not start. We push-started it for the first of many, many times that day. After another three or four kilometers we were climbing up the mountain by the river and the waterfalls where the wrecked hydroelectric plant was located. We were close to some gold mines, and just at that moment Tom's bike stopped running again. We didn't know it then, but this would be the pattern for our day of travel. That time it was a battery problem, which took Tom an hour to fix, but fix it he did. The man is a wizard with motorbike repairs. Reaching the top of "Telephone Hill" we took a short break, followed by again push-starting Tom's bike.

Our trip looked a lot like this: go, stop, mud pits, brackish water, river crossings on logs, bridges out or just a few logs remaining, in which case we had to ford the muck. The first twenty-seven kilometers—fifteen miles or so—took three hours, and I counted at least twenty-five stops for *piki piki* problems, or because someone's bike had fallen into a mud pit, or just to clear some obstacle or obstruction across the path. After those three hours, we were once again covered in red mud and smelly water.

Somewhere in the first twenty-seven kilometers the rear tire of my *piki piki* went flat. We pumped it up with a hand pump three times, but only made it about one kilometer each time. Finally, we reached Lugungu village, where Berean Mission has a church. This was also a major "cross paths," as well. The local pastor had already gone toward our upcoming Leadership seminar. We still had forty-two kilometers—twenty-five miles—to go, and we had been on the path for over three hours, and we had a flat tire on my *piki piki*.

Ramazani walked the village looking for a replacement inner tube for the tire. We *Bazungu* (white people) slept on the ground under a big tree while we waited. Rather crazy, as we were surrounded by local African kids and adults, watching us sleep. The nap came slowly, since the crowd also cut off any breeze that may have been blowing. Ramazani found an inner tube somewhere (don't ask) and fixed his rear tire. We took off again on our last forty-two kilometers.

The path became less and less passable. There were ten-foot jungle walls on both sides, more brackish water, rivers, mud pits, and more. Stop and go, more mud, more falling. We pushed ahead with our goal always in mind. Around 4:00 p.m. we hit a stretch of path we used to call the *Katangila*, which I recognized immediately. This stretch was in a clearing and one could see station Katanti up ahead. As kids we used to watch this stretch from our porch, to identify who might be coming to our station. In later years it was also used as an airstrip. It was now overgrown, with just a path to ride on; however, it meant we were only two or three kilometers from our destination. My heart started to really pound with excitement.

At about 4:30 p.m. we turned and climbed the hill to station Katanti. By then word of our coming had circulated, and the path was lined with Africans—singing, shouting our names, and crying with joy. I got goosebumps the size of M&Ms this time. I was home! What a jubilant welcome we received. We rode through the crowd and stopped in front of Jansen Hall, my home. Immediately, we were mobbed by happy Africans.

SaSimon, who had been an evangelist and teacher with my dad all those years, was waiting with close to one hundred people.

We dismounted from our bikes and were engulfed in a sea of hugs and cries of joy. SaSimon led a prayer of thanks for our arrival. The crowd was also so happy to see Tom and Kathy, since it had been three years since they had been able to go downcountry. They asked me to say a few words to the gathering, with Tom interpreting.

Words could not express what was in their hearts and our hearts. I greeted them and gave them some Kilega (African) proverbs that I could remember. Such joy, laughter, and carrying on.

We were then led into "my house," a brick house with aluminum roofing, built around 1950. It was still standing, but crumbling and falling apart. Four African families lived in my house, as guards. I wandered slowly from room to room downstairs, taking pictures of each room. I felt overwhelmed and just let the memories flow over me, later taking my bag and backpack to what used to be the pantry.

The Africans had made up a primitive low wooden platform, a bed under a mosquito net, and a very rickety shelf on one wall. My "home away from home." Tom and Kathy got the same things in a larger room my family used to call the guest room. The dining room still had the same old wooden tables that were there in the 1950s. When I was a kid we used the two wooden tables for ping-pong tables! So good to see them again. Now we were going to eat on those very same tables—unreal.

Another "moment" for me was the sight of Mother's old Singer sewing machine, still sitting in that dining room! Apparently the rebels did not know what to do with a sewing machine.

Next I walked slowly around outside the house, as it was getting on toward dusk. The front steps and porch were still there, although they too were crumbling. (I made a decision to wait to climb to the upstairs floor until the next day, when I would have better light.) Then I walked the station with Kathy and Elena, SaSimon's daughter. His wife is deceased, so his family now lives with him in a different mission brick-and-tin house.

My "original" mud-and-stick-and thatch house from 1950 was gone, and the jungle had taken over that place completely. In fact, the only two houses still standing from that era were the brick and tin-roofed ones. The cement-block schoolhouse where I attended school was still there, but the church was gone, burned and pillaged by soldiers more than once. I found what was a carpentry shop, and the softball field we played on all those years earlier. (The Africans could not pronounce "baseball," so they called the game "Fayball.") That was because they heard us yelling, "Fair ball!"

I went down two jungle trails where I used to go *Kubisindi* (to hunt squirrels) back in the day. What fun, reliving the old days.

That same day many people, older and younger, came to see us. Some had worked with my parents, while others were sons or daughters of the "old timers," and even grandchildren. What a reunion. A group of Africans came to serenade us, as a welcome. The group sang and danced nine songs.

When the singers left, we took turns taking "spit baths" in what was left of the old concrete tub, in what was the girls' bathroom in the dormitory back in the day. Although the tub still drained, we saved the used "gray water" for more people waiting to bathe. If you get my drift, the last person got quite used water for their spit bath. To be fair we rotated baths each day. The water had to be hauled in pails, and it was cold too, but oh well, better than nothing.

Each of us had brought along one bath towel and one wash-cloth. The toilet had long ago been torn out of the girls' bathroom, but across the hall was the boys' bathroom, which was used for the dormitory boys back in the day. Its toilet was still there, and was flushable with a bucket of water. Don't ask where the flush goes—I had no idea, and really didn't want to know. Oh, you wanted a toilet seat? Sorry, that had been missing for decades, so we just sat on the porcelain.

Word had come to us that a local official wanted to see our passports and visas. Specifically he had asked for mine, so we sent them all down a few kilometers for him to inspect them. Once again our documents were in the hands of strangers. The same man sent word back that Tom and Kathy's papers were fine, but he needed to see me, since he said mine were not fine. Baloney! Mine were just fine too. We decided to wait until the next day to go see him.

Dinner was brought to us by the four "bodyguards" at 8:00 p.m. We had rice, greens, chicken in palm oil, and *bugadi* (made from a local root). It all tasted wonderful, just like I remembered from my youth. Yummy.

Our beds were made up with mosquito nets. One of the NGOs had given every African three mosquito nets a year or so prior. Probably this was to fight malaria. You wanted a pillow? Get serious, *This is Africa!*

What a joy-filled day. We sang a song with the Africans, prayed, and then went to bed. My first day back at my house—and getting to sleep in it, as well! I fell asleep dreaming of the old days.

CHAPTER THIRTY-FOUR
CONGO STRIKES AGAIN

I AWOKE VERY early the next day, feeling like someone or something was in my room. It turned out to be nothing, just wild dreams.

Breakfast was dry cereal we had brought with us, powdered milk, freshly boiled water, scrambled eggs from local chickens, and fresh papaya. The scrambled eggs in Africa are very, very greasy and may contain things other than scrambled eggs! My son-in-law Bob Veach says, "God, I'm putting this down into my stomach now You please keep it down!" Amen to that.

Going upstairs in my house, I walked through the boys' dorm room, the girls' dorm room, and my parents' room. In the hallway between these rooms I found that an African was living on the floor, sleeping on an old set of springs, and no mattress.

The bedrooms upstairs had been pillaged and looted many times, as well. One interesting thing stood out. The soldiers had pulled out all the electrical wiring in every room. The story was that someone had fooled the soldiers into believing we missionaries used the wiring to hook up to the Internet and plot against the soldiers and Africans. Never mind there was no Internet. It was just crazy what they would believe.

Tom and I took his *piki piki* to Mapimu, down the path, to see the "Minister of Something" regarding our papers, particularly my visa. The official had claimed it wasn't "authentic." The path wound about six or seven kilometers through the jungle. About three-fourths of the way there, Tom's *piki piki* ran out of gasoline. It wouldn't run at all on the "reserve" tank, due to dirt in that tank. We left the *piki piki* on the path in the forest, took the key, and walked the remaining way to see the official. He was busy, so we were kept waiting for a while before we could see him. This too is part of the hazing and not unusual.

Finally, we got in to see him. He turned out to be a very pleasant man, exchanging greetings with us and asking our purpose for visiting "his area." We did not bring up the paperwork issues, and neither did he. After a few minutes, he said we were fine, and we could go. Later we found out he already had local issues that morning. A crowd had gathered to demonstrate against him for his handling of forest and garden rights.

We also heard via the grapevine that someone on a radio station was announcing that we were returning to Katanti with a sinister purpose. According to the erroneous report, we would be going to the spot of the original mission built in 1938. At that spot, Tom, Kathy, and I would dig up "buried treasure" our parents and others had left in the jungle in 1960! Nothing could have been further from the truth. As a matter of fact, none of us even knew for certain how to find the original mission spot. Furthermore, we did know for certain there was no buried treasure nor anything else buried there. This did give us some pause, however, that our mission was being publicly misrepresented in such a way. All we could do was to pray about it and trust God for our safety.

Tom and I put out the word in the "Official of Something's" village that we needed to buy some gasoline, again in used plastic drink containers. One man went off to find some for us. In the meantime, we looked at the items in a ramshackle path-side "store." Actually we needed a pair of scissors, but they had everything but scissors. One could buy a solar panel, colorful fabric, etc., but no scissors. Go figure! While we looked in that place, some men were using a rudimentary kitchen scale to weigh tablespoons of gold dust, right out in public. Amazing!

After a while we walked back to where we had left the *piki piki* and waited there for the man to bring us some gasoline. He arrived after a long wait, and after we put the gas into the bike, it roared to life. When we arrived back at station Katanti, our self-described "bodyguards" were doing our laundry by hand behind the house. They hung it out to dry in the small attic of my house.

That day we needed to start the Leadership seminar. Tom taught "Five Major Doctrines of the Bible"; Kathy "The Full Armor of God"; one of the pastors taught "Corinthians"; and the senior pastor did a daily wrap-up. About thirty-five to forty men came to the seminar and maybe twenty women. It was the season to be burning the forest patches for gardens. That meant the women could come only in the afternoons, due to their garden workloads. At the end of day one, I was asked to speak to the crowd. I told them about my family, my extended family, and my faith in God. Tom interpreted my message. While the group really appreciated my words, I was pretty emotional. No matter, I wouldn't have missed the opportunity for the world.

As the seminar ended, Kathy and I went with an African woman to her village to look for our former cook, Kibekyangabo. Kibe had been our cook for years and years in Congo DRC and was now in his nineties, but his health was failing. What excitement and joy as we found him in a wicker chair in front of his small house, waiting to greet me. We had a great chat, with Kathy interpreting. On our way back up the hill to my house, we passed five or six graves of the pastors who had worked with Dad in the 1950s and 1960s.

I had a huge event that day. My two best childhood friends, Yoanne and Yosef, came to see me. So good to see them again! We laughed, cried, talked, and basically had a grand old time, as guys everywhere love to do. These were the same guys with whom I had hunted squirrels and had many other great times between 1950 and 1960. What a good reunion, in spite of high humidity and very hot weather.

Just before I went to bed, the local pastor came to talk with Tom. He asked Tom to cut the seminar from five days to three days. The reasons: 1) They were running out of food to feed the delegates; and 2) They needed to be working in their gardens as much as possible. Tom explained the difficulties of cutting it short and promised a decision the next day.

That evening we went to bed under a beautiful three-quarter moon, shining right in where the window was supposed to be. I noticed that I didn't hear wildlife calling in the forest at night like I had back in the day. I couldn't wait to see what the next morning would bring.

CHAPTER THIRTY-FIVE
ARRIVED

I GOT UP early for a devotion time and to wash up a bit. The day dawned clear and bright. Kasimbula (house guard) said to me, "It is so strange you don't remember how to speak our language." Nice compliment, and I wished it would return to my brain too.

We met a man whose father was Lutilitili. His dad is deceased, but back in the day his dad used to dress up like a witchdoctor and scare the living daylights out of us missionary kids. He would also make a sound like a leopard and around that sound he would wrap these words: "I want to eat a missionary kid!" I shared good memories with his son.

The previous night I had found a group of Africans sitting around an outdoor fire, watching a DVD of Tanzanian singers on a Chinese battery-operated player called a KOBY. One of the non-governmental organizations had given them out in the past year. The people kept it charged with a solar panel—amazing how they have adapted to solar, when they have nothing.

At Tom's suggestion I left the seminar at the first break of the day and went walking all around station Katanti. Once again I just let the memories flood over me. First I went down several jungle trails where I used to hunt and trap squirrels. The gasoline electric generator and water pump were long gone. I found another car rusting in the jungle, then sat with Elena, SaSimon's daughter, on a bench under some bamboo. She showed me a book called *Major Doctrines of the Bible* that she had kept all these years. The book was translated by Dad (Ernest Green), C. Neal VanderPloeg, and an African named Sansago Jerome. What a wonderful moment to look through the book! She also showed me old, old pictures she had treasured over the years. They were of the earliest and later missionaries to come to this area. Somehow she had kept these hidden from soldiers all these years. Since she did not want to part with her treasures at any price, I took good pictures of her prints.

As I walked through the jungle and the tall grass, I used a six foot pole to swing back and forth in front of me. This was to ward off any snakes that might be hiding in the brush. Trekking behind my house I found three large guava trees, which had been planted back in the day. Back then, we kids would carve our initials into the bark, inside a carved heart, with the initials of our current girlfriend carved with ours. I wanted to do something extra special, so I carved a heart with my initials and my wife's initials into the tree, as a tribute to Carol Green, my wife of forty-five years, who is now in Heaven. The carving, a special moment for me, will be there for years and years to come, just as the older carvings are still there fifty-two years later.

I investigated the carpentry shop, schoolhouse, baseball field, and even found the tennis court we kids had created with picks and shovels long ago! That was where I learned to play tennis in the 1950s, on a clay, homemade court. The court had been taken over by the jungle.

That night a huge rainstorm with lightning and thunder and high winds came up. We sat on the porch and watched it come across the valley. During the storm, we caught fresh rainwater from the roof, for spit baths and cooking. Tom had made the decision to end the seminar on Thursday, so we hoped to get going out of the interior by Friday or Saturday. Friday was the day most planes went from the regional post back to Bukavu. If we could not make it out on Friday, things could get complicated for my schedule of flights back to the United States.

One pastor who had walked thirty miles on crutches asked me if I had any newspapers I could give him to practice his English. I did not, but I gave him a *Reader's Digest*. He was very grateful. We also had to be very careful to each lock our individual rooms in my house, as well as the doors into the house, as things would "disappear" if we didn't. Each of us had our own keys.

Also, we discovered evidence of rats in the house, as well as snakes. What else is new?! Tom had perfected a great way to get rid of rats crawling on his bed, while he was in it. He carefully slid his foot underneath the sheet, as close as possible to the rat. Then he would give it a mighty kick through the sheet, watching the rat fly against the wall! Worked for me.

In talking with Tom and the Africans, I discovered that some missionaries had actually lived at station Katanti off and on until the early 1990s. However, they were in and out with evacuations, and fleeing the soldiers' uprisings. When the soldiers came through, they killed any men who had not fled, took the boys as child soldiers, took the young girls for sex slaves, and then raped the women. After that, they burned everything they could. Crazy. These practices continue to this day.

In the middle of that week I awoke to a big surprise. Twenty to twenty-five men came to give me a special gift. With much work and fanfare they pulled and pushed a black goat roped behind them. Someone made a *big* speech and presentation: "Roger (*Banuamazi*), we know that our lives here are very hard. You now see the struggle we have living here. Even though we don't have much, we got together and want to give you a very small token of our appreciation for your coming to visit with us. So we are presenting you with this black goat."

Immediately recognizing that this represented a sacrifice of several days' meals for the Africans, I thanked them profusely. All the while I was wondering what on earth I would do with a live goat. (Sure couldn't put him in the overhead bin on KLM airlines.) My fears were instantly allayed when the spokesman said, "We are so happy you came back to visit us after all these years. We know your travels were *hard*, on poor airplanes, *piki pikis*, old taxis, and even on foot. So with your permission we want to slaughter the goat and have a goat banquet feast with all the missionaries and all the seminar attendees." With some relief and some trepidation I immediately agreed. Relief that my "goat problem" was resolved, and trepidation in that the last time I had eaten goat in Africa it did not sit well with my stomach. However, this was such a touching tribute, I knew I'd eat it. At mealtime, I carefully selected one small portion of something I knew was goat *meat*, not the entrails, which the Africans see as a delicacy. Ugh.

Four pastors' delegates came to speak with me after the banquet, which we all ate, packed into my house. Basically the four pastors were asking me to stay with them in the jungle for two or three months, so they could refresh my Kilega, the local language. That request had sneaked up on me. I had to tell them I had passage booked back to the U.S., and we would have to see what God had in mind for the Africans.

SaSimon, formerly Dad's right-hand man and now in his nineties, made his own passionate appeal: "Now, to all three of you, we are so happy you came through all your troubles to visit us and present a Leadership seminar and eat a banquet with us. Now that you three have blazed a trail into the jungle, we want all the other missionaries' kids to come see us, as well. In fact, you come again and bring them with you." To me he added a personal touch: "Roger, *Banuamazi*, we are saddened by the loss of your wife, Carol, and your daughter, Nicole, but the fact that you came through the jungle proves it can be done."

The Katanti seminar became an appeal to Tom, Kathy, and me to get our kids to sign up to be full-time missionaries in Congo DRC. Specifically, the Africans wanted our kids to come work with them at stations Katanti and Katshungu. The Africans now saw us as trailblazers, similar to the six original Berean missionaries in 1938. Tom did his best to explain the logistics, dangers, changes, and issues, to which SaSimon poured on more pressure, saying: "Your parents started this work in 1938, and built this work here. You can see it still going on to this day. We beg of you not to forget us, and to bring us new missionaries to live among us."

Tom gave them a kind response, which they sort of accepted. In part, Tom said this: "We are trying to set up a mission office in

America as a liaison to you here in Congo DRC. That has not yet been accomplished, but may be in the future." In addition, Tom told them, "Asking young couples in America to commit to a lawless land, with no roads, no electricity, little water, scarce food, mud-and-stick houses, and danger all around for them and their families would be a hard sell to those American young couples."

After all kinds of *talk, talk, talk, blah, blah, blah*, I gave my response: "Thank you so much for inviting me to come visit you, dear ones. I have heard what you said, and right now my head is spinning. One thing I can say is that I am praying for you daily, and I will not forget you. God bless you each one!" We all sang "When They Ring Those Golden Bells for You and Me." Tom led us in prayer, and the Katanti Leadership seminar was officially concluded.

Later that night, Tom and I sat outside, exchanging snake stories into the night. We really wanted to start our journey back to Bukavu the next day, but suffice it to say, once again, *This is Africa!* There were still many issues to be negotiated and talked to death before we could leave. As I said before, my chances of getting to Bukavu to connect with my flights to the U.S. seemed to be getting more and more remote. As they always said in Congo, *Tutaona*, which means "we shall see," or as I say, "stay tuned."

The final day of our time at Katanti was a Friday. We were ready to leave, but it was not to be. The African family living in the room across from mine awoke very early, like about 4:00 a.m. Why? I had no idea, but sleep was at a premium after they started talking and moving about. I found out later they had gotten up early because they thought we were going to leave and wanted to give us a good start on our way.

Tom and Kathy were already up too, dividing up supplies for the schools at four Berean mission stations. We had breakfast later than usual, since our rations were by then quite low. In addition, we had to get our body weights down as far as possible, since Tom's *piki piki* still was not running properly. The lighter, the better.

The "bicycle guy" came so we loaded what was left on his bike for transport to Shabunda, thirty-five miles away. During the process we discovered the other man who was supposed to speak with *piki piki* drivers to come and get us had failed to do so. At that point all we could do was put out the word for drivers, and again commit our journey to God.

Since we were not going anywhere any time soon, I went to take a spit bath. One of the houseboys had carried water for me all the way from down the hill in the village below. He was pressing me to use that water, and not to waste it. I took a spit bath and washed my hair.

Many of the Berean churches' delegates were hanging around, hoping for something from us. (Often when heading out from stops in Africa, there were things we needed to leave behind.) We did give them some gently used batteries, shoes, and clothes. And one of my best friends from yesteryear, Yoanne, came back again. He brought us fresh papaya and fresh bananas, just what we needed for breakfast and snacks. With heavy hearts we bid each other goodbye.

Later that day, while we waited for some drivers to show up, the pastors wanted to have a meeting with Tom and Kathy. I tried to take a nap, but my brain was too rattled. So I just laid down to rest and pray. After their meeting, the three of us walked about five kilometers to see Sansago and his wife, NyaDenga—Yoanne's parents. This couple had twelve kids—one has died—and they are great-grandparents now. Sansago still remembered a little English he had learned years ago. On our long, hot walk back up the hill, we saw another friend, KingNgombe. He just shouted when he saw me.

Late that day, two *piki piki* drivers came with five liters of gasoline for Tom's motorbike. Things were picking up for leaving. The two men went to find a new valve stem for the rear tire on Tom's *piki piki*, and then returned to install the tube. It appeared that we'd be able to leave the next day bright and early to begin my trip home. As we made preparations, SaSimon came and sat on my porch, to sit and cry. He was distraught that our time was almost over, as if to say wait, *This is Africa*! But it turned out we weren't going anywhere the next day either.

For the third time we thought we were ready to go, while we waited and waited for things to come together so we could leave. Yes, I was getting impatient, but I reminded myself that "God is good all the time" and that *This is Africa*.

While we waited, Kathy and I walked to the clinic and dispensary. On the way we saw both an old building and a newer one. The old maternity building was where Mother delivered literally thousands of babies back in the 1950s. It had always seemed she was called down there in the middle of the night, never during daylight hours. There are a plethora of stories about Mother's work delivering babies, helping injured folks, sick folks, people attacked by crocodiles, snakebite victims, and those afflicted with many other maladies. Often she had to do the work of a medical doctor, even though she was trained as a registered nurse. You see, there was no medical doctor.

Another pastor came to talk to us, saying, "We are happy when you come to see us but it is so hard to see you leave!" This was the same pastor who told us in 1960 that we had to leave the Congo or be killed.

Late during that afternoon of interminable waiting, we heard a message wafting over the air from an African "talking drum." Since the African dialect is a language of different tones, high and low, the drum is a very good way to communicate across long distances. The drumbeats were coming from Masegesege, a village perhaps six kilometers away. The message was that someone's son had been killed in North Kivu Province in the rebellion and uprising there. Word had just reached the village, and the family was in mourning.

Some African boys picked and brought us fifteen fresh guava fruits. I eagerly ate one. Back in the day we used to make wonderful jelly and jam from the guava fruit. We had probably planted that guava tree fifty years earlier, and it was still bearing.

With our time crawling slowly along, Tom again answered questions from African pastors. In one and a half hours he covered a lot of ground:

1) Economics and foreign investors: The pastors asked why no one wants to invest millions of dollars into the Congo DRC. Tom replied that foreign investment is at a near standstill because of the security and safety issues in Congo DRC. Any investments that would be used to build infrastructure or housing could be lost if the soldiers came to loot and burn stuff down. In addition, the Western and Eastern powers can get minerals from the Congo DRC on the black market for cheap. So why would those countries want to invest when they can already get the minerals cheaply?

2) Obama's presidency: The Africans could not understand why President Obama had done nothing, in their eyes, to promote Africa on the world stage. We told them that was a good question, but that

many of the same factors that affect the lack of investment hold true for promoting African countries.

3) Rampant divorce in America: Tom discussed the African dowry system for a marriage. In Congo DRC once a father has been paid a dowry price for his daughter to marry a man, she is that man's wife forever. They do not understand quick legal divorces in the U.S., nor gay rights issues, which make no sense to them.

4) Same-sex marriage: These pastors just could not get their heads around the concept of legal or illegal same-sex marriages in the U.S. Their question was really very simple: "If there is a same-sex marriage, which one is the man?" In Congo DRC, the man in a marriage is very much in control and in charge. So, for them, this was a serious question on a serious issue. Tom did his best to attempt to explain the changes in American politics and issues over the years.

After those heavy discussions were finished, a ten-year-old African boy named Debaba was shooting birds with his slingshot, as I used to do. I watched as he killed four hummingbirds in short order. He and his family would bake them to eat, although there isn't much meat on a hummingbird. Back in the day I could shoot like Debaba, but not anymore.

We ate a meager dinner at my old table, as our rations were very low. The plan was now made. We would get up before dawn, and at first light we would leave on the first leg of our safari. The *piki piki* trip to Shabunda could be three hours or six hours, one could never tell. We hoped and prayed to be able to catch a cargo plane midafternoon, to get back to Bukavu the same day. Since planes only flew during daytime hours, it was imperative to get going early. Of course, that would be if they flew at all.

Return airfare was only $159 U.S. each, compared to the $300 U.S. we had paid coming downcountry. Since the planes flew back mostly empty, the fare was cheaper. We were happy about that. So we went to bed thankful that everything was in place once again for an early departure. Or . . . so . . . we . . . thought!

CHAPTER THIRTY-SIX

LEAVING DOWNCOUNTRY CONGO DRC

I WAS AWAKENED two times in the night by wind, thunder, and lightning! Rain was blowing into my bedroom space through what used to be a window. We really needed to get going that coming morning, and God did provide, as you will see.

We got up at 5:30 a.m. and had a few bites to eat at 6:00 a.m. Guess what? The sun was already out and drying the paths for us. Then a new problem surfaced—*This is Africa!* Tom's *piki piki's* rear tire was now flat as a pancake! Our *piki piki* drivers went out looking for an inner tube and some backup gasoline. (They had not brought all the stockpiled gas, and we could run out.) The guys found gasoline, but no inner tube. We prayed, pumped up Tom's tire, and waited an hour. Praise God, the tire held its air.

So around 10:00 a.m., after emotional goodbyes to the Africans, we started out. I cried all the way down the hilly path on station Katanti, on the back of my *piki piki*. This time my driver was a really nice young man, and a good driver. I found he knew three words of English: yes, no, and okay. As expected, the path was a mess, from the night's rainstorm. We had to ride very carefully, with many dismounts and pushing the bikes. The rear tire on Tom's bike held until we were about four kilometers short of our first goal, a place called Mapunga. We stopped in that village to regroup. Over the course of an hour and a half, the Africans took off the back wheel and tire, and patched holes in two places. (Now, why didn't these guys tell us they had patches and glue in their toolkits before starting out?! On the other hand, we were surely glad they did.)

The next phase of our journey was the path from Mapunda to Lugungu. This stretch of the path was an amazingly difficult passage. Trees down, many, many rivers, many, many culverts, very overgrown, you get the idea. Very slow-going, until we finally reached the crossroads of Lugungu around 2:30 p.m. We rested a bit, then headed out on the final leg to Shabunda regional center and airstrip.

Just six miles short of Shabunda, Tom's *piki piki's* rear tire went flat again. We negotiated with some villagers to push Tom's bike to our destination, and we would walk. Seeing we were out of options, their price was exorbitant. They wouldn't negotiate the price, so we went to Plan B. Tom went ahead with one of the good *piki pikis* while Kathy and I stayed behind with the broken-down bike. Tom would return or send two *piki pikis* to pick us up before nightfall. Kathy and I sat under the thatched-roof-covered meeting place in the center of the village, waiting and watching over Tom's bike.

While we were waiting, a plane—our plane—flew right over us on its way to the airstrip at Shabunda. Our hopes of making it onto that plane that same day were dwindling to zero. I kept repeating, "God is good all the time." Now our hope became to get a plane to Bukavu the next day, Sunday.

Meanwhile, back at the village a crowd had gathered. A couple of the people knew passable English, and with Kathy's Swahili and my Kilega we were able to converse with them. Two of them said they were the Catholic priests from the village. But one of them said to me, "If you will take me with you to America and give me a job, I'll gladly switch to your religion." The two men were drinking palm wine or palm whiskey, made locally from the fruit of palm trees, the palm nuts. They "dared" me to taste it, so I did. I thought to myself, "Now I won't need an airplane to fly!"

Kathy and I waited over two hours and began to get a little nervous as the afternoon wore on toward darkness. An old man came to the meeting place to see us. He couldn't keep his eyes off Kathy, which soon grew to be an uncomfortable situation, especially since he had been drinking and was, in fact, drunk! At one point he asked Kathy to "go home and pray with me and cook for me." In Congo DRC that last phrase means he wanted to make her his wife. Of course, she refused. This was met with more bravado and jabbering on his part. Later, he again asked Kathy to go home with him and cook for him. So we decided we needed to take some action of our own. We pushed the bad *piki piki* up the hill and away from the meeting place. Thankfully the crowd, including the old drunk man, did not follow us. Some thirty-five minutes later, two *piki pikis* returned to pick us up. We left Tom's *piki piki*, telling the Africans that someone would come for it the next day. For good measure, we took the keys with us.

In the dusk of late afternoon, the eight-kilometer ride to Shabunda was uneventful, until we approached a crowded crossroads filled with people, including police and soldiers. I noticed all of them were standing very still, and it looked menacing. My driver stopped and very quietly told me to get off the bike, and to stand very still. My thoughts were "this can't be good" and "this may not end well," so I stood there praying for safety, all the while standing still. After a time, a whistle blew and everyone began to move again. My driver told me to get back on the *piki piki*. Off we went to our destination, the Shabunda pastor's house. We found out later that every day at sundown they take down the Congo DRC flag. Everyone who happens to be nearby must stand at alert until the all-clear.

Shabunda was teeming with people, and I didn't feel completely safe at that moment. I couldn't help but flash back to 1960, when we had fled this place. We also discovered from the pastors that because of the erroneous report on the radio that we missionaries were coming back to dig up buried treasure, a number of pastors had put their own lives at risk for us. They had gone on the radio themselves to tell listeners that the story was a hoax. In 1960 the African pastors had stood in front of the soldiers' guns, blocking the soldiers from shooting my parents and others during our evacuation.

The pastor's wife had boiled drinking water for us, and we were also able to take a spit bath for our weary, hot, tired bodies. I was given a small room with a mat bed, and Tom and Kathy had a larger room with the same thing. The bathroom, a hole in the ground, was down an outside path, through the pouring rain, which had started up again. Shortly after a meal of the same things we had been eating for two weeks, I went to bed. I was beat. Sleep came quickly. The rain leaked through the thatch roof onto my bed, but I didn't care. The first leg of our trip out of the bush was completed, and I could hardly wait for what may happen the next day, a Sunday . . .

CHAPTER THIRTY-SEVEN
FLIGHT FROM HELL

SUNDAY AT SHABUNDA regional center dawned partly cloudy, after it had rained hard two times during the night. The roof over my bed in the African pastor's house leaked, and the bed got wet down at the bottom. I just slept with my legs drawn up, in the dry part of the bed. I finally got up around 6:45 a.m. and waited for the rest to arise. Breakfast never came, so we went back to the good old Power Bars I had toted in my backpack. Those were the last ones I had.

On Sundays there were not as many cargo flights out of down-country, so we would take our chances, hoping to catch a ride on one, if one came. While we waited, the Africans brought us some boiled water, from which we made instant coffee.

Tom showed some of the pastors and others watches and glasses he had brought from the U.S. Since we were going to take any cargo flight out, we had decided not to attend church, but rather to await word from the airstrip and be ready to go at a moment's notice. Suddenly we got word that there was a cargo plane leaving for Bukavu in fifteen minutes and the pastor would come to get us to the airstrip. We rushed around like chickens with our heads cut off to get our things ready to go.

As usual our escort took thirty-five minutes to show up. Finally, we started our two-mile walk to the airstrip. Upon arrival we noticed the Let M-410 plane was the same one that had broken down the night before at the airstrip! This would be our plane for the day. Everyone connected with the cargo flight assured us that even though the plane had broken down just the night before, it was fixed and ready to fly. Well, what could we do? The plane was already loaded with cargo and a few people, and just waiting for us to show up. Thank God they waited. Immediately, we boarded without any paperwork snafus, such as when we had been held and hassled a week before. We all climbed up a vertical rope ladder, which was barely big enough for each of us to climb up and into the plane's belly. Praise the Lord, we were going back to Bukavu.

Once again, the inside had been totally stripped and made into a cargo plane. There were two African pilots in the front, their seats open to the back. No bulkhead nor door between them and us. No crew, only the pilots. Someone had left four seats bolted to the floor of this plane, in the middle of the cargo area. Those four seats were occupied by Congolese soldiers. The pilot came and told the soldiers they would have to give up their seats and let the "paying" passengers have them. The soldiers were not happy. We three agreed we'd rather have the soldiers "happy" and just let them have the seats—but the African pilot wouldn't hear of that. So the soldiers sat on the cargo, looking quite upset. That was all we needed—unhappy soldiers at the beginning of another adventure flight over the jungle.

A young African businessman and student sat beside Kathy, and couldn't keep his eyes off of her, making her uncomfortable. Tom and I sat right behind them. Oh, you wanted seat belts? As they say in New York, "Fugetaboutit!" Some barrels of something were rolling freely around toward the back of the plane. Of course, the bathroom had also been torn out of the plane.

As we silently prayed, the pilot started each of the twin turbo-prop engines. Although they insisted everything was fine with the plane, I was comforted to notice he ran each engine up to speed for longer than usual. Satisfied, he released the brakes, and we lumbered down the grassy dirt runway. Amid much noise, the plane lifted off, and against all odds, the crazy thing actually flew. Keeping close watch on the unhappy moods of the soldiers glaring at us, we headed toward our destination. We really had no idea what might happen. Thank God our trip to Bukavu was only forty-five minutes, punctuated by a good, soft landing.. (When flying over the jungle interior, one can see wreckages of numerous flights that went down, even one where a missionary died.)

At Kavumu airport, outside of Bukavu, immigration was a breeze. It felt good to be back in a semblance of "civilization." Upon arriving at Tom's apartment, we discovered that once again there was no water and no electricity. And our emergency batteries were getting quite low on juice. We were, however, able to charge our international phones from those batteries. Then we could make calls and send texts to let our loved ones know we were okay. Remember, we had been out of communication range for nine days.

Since it was Tom's birthday, we went to a restaurant within walking distance of the apartment. It was clean, and we shared a nice meal there. Tom had a fish entrée—and I mean the whole fish, head and all. Kathy had tilapia and potato fries. I had pork chops and fries. I treated Tom and Kathy for his birthday, and it was only $45 U.S.

After returning home, we all three took much-awaited long naps, until late evening. Then we ate leftovers for dinner. A big windy, noisy storm came up after dinner. We had to batten down everything against the wind.

Even after the long afternoon nap I was "wasted," so I went to bed around 8:00 p.m. I loved hearing the rain on the metal roof, especially now that we were out of downcountry. In just two days, I would be retracing my steps and flights from Congo DRC through Rwanda, Uganda, Amsterdam, Detroit, and into Dallas and the good old U.S.A. Meanwhile I started giving away clothes, shoes, jackets, and a suitcase, with a goal of getting everything into just my carry-on backpack.

CHAPTER THIRTY-EIGHT
MOST "HELLACIOUS ROAD"

ON THIS FIRST day back from downcountry, a man came to the apartment in Bukavu to take Kathy and me to see an orphanage called Kaziba Orphanage and Hospital. Before going downcountry we had met the director, Mr. Flavien. I particularly wanted to see his operation and compare it with Buyamba, Inc. in Uganda, with which I'm affiliated.

Kaziba Orphanage and Hospital has about sixty kids in residence, from newborns to six years of age. Once they reach age six the orphanage does all it can to get the kids placed for adoption, or into the homes of foster parents. It was started many years ago by Norwegian missionaries and is now run entirely by Africans. Many supporters and/or sponsors of these kids are European or American, and other nationalities. The kids are most often from families torn apart by the many, many years of uprisings, wars, AIDS/HIV, or just "unwanted" babies. How sad!

The "road" we took in Flavien's Land Cruiser was the *worst* "road" I'd seen in Africa, anywhere, any time! (The jungle paths we had ridden through on *piki pikis* were indeed worse, but they were just paths.) The Kaziba road was mountainous, rocky, and bone-jarring, and had mudslides, one-way hairpin turns, and took us four hours to go thirty-five kilometers. Also, this "road" was used by huge trucks carrying red mud-bricks back to town, most often with another load of many, many people clinging to the top of the load of bricks. I wondered how in the world they could hold on to stay on top of the load. But hold onto the load they did. Some of those trucks were two stories high. Another hazard was the groups of people on foot, walking that road with heavy loads on their backs. Our driver was cautious, and I don't think we ever traveled above five miles per hour. The driver would sound the horn every time we approached a "blind" corner, of which there were about five million.

Along the way, we came across two or three huge trucks that had slid off the road and into the ditch. On one occasion, a truck was blocking the road ahead. Another truck came to offload the first one, and since we were on a one-way stretch of road, we had to back up a ways before we could fit past. I thought, *What's that sheer drop on our side of the road? Never mind.* To sum up that road, by the time we reached Kaziba we had broken a shock absorber on the Land Cruiser. Believe it or not, in the middle of nowhere someone fixed it while we toured the orphanage and hospital.

The Kaziba post was very well organized, had three African doctors and the hospital, and had good places for the kids to live. It also had a huge, working hydroelectric plant built on a river that has survived since the Norwegians built this post many, many years before the rebellions. It boggled my mind that the soldiers had not come to destroy it as they had the same kind of power plant in the jungle. Probably it is a testimony to how remote this post is, and the difficulties of driving there.

We toured the orphanage and hospital and fell in love with the sixty little kids living there. One tiny baby had been brought to the orphanage early that same morning. The child had been born that very night, and its mother had passed away during childbirth. Kaziba took the child in so it would not be killed or tossed away to die. My heart went out to that little tyke.

After lunch with Flavien and his wife, we started back up the long, long road. We needed to get back to Bukavu before darkness fell. Several sections of the road went through places where it would not be good to be out in a car after dark, especially for white people. Along the way we came upon two of the huge trucks that had just had a minor scrape (fender bender). A mob of young African men had climbed down off their perches on top of the loads and were yelling and fighting each other hand and fist, or with poles and sticks—all over the fact that someone had broken seven bottles of whatever kind of beer was being transported. Flavien knew some of the combatants and stopped to see if he could broker some kind of "peace deal." After a bit, and at the urging of Kathy and me, he gave up and drove away. We breathed a sigh of relief.

We reached some of the most dangerous areas of Bukavu just at 5:30 p.m., as twilight descended. The roads were again teeming with thousands of people on foot, so thick that cars could barely make it through the mobs. Finally we arrived at Tom and Kathy's apartment at 6:15 p.m., almost full dark. Kathy and I were absolutely beat, from the pummeling we took on the road and from some of the dangerous moments we had survived, as well. For dinner we gratefully devoured peanut butter and jelly sandwiches.

As if knowing I would soon be leaving Congo DRC, a man from the Berean church came to visit me again. He brought me a passionate plea again. He begged me to tell my kids and grandkids back in the States to come to Congo DRC and work with them. As before, I carefully told him that would be a "tough sell."

The next day would be my last full day before flying out of Africa. I went to bed about 9:00 p.m., wondering if the next day would bring as many surprises as what I'd been through up until now. I was certainly caught up on my "surprise quota." As I went to sleep, I was happy to be helping out several Africans by leaving them my "gently used" stuff. I didn't think about it for long, as sweet sleep came quickly.

My final full day in Congo DRC, I awoke at 6:45 a.m., took a spit bath, and shaved. In Bukavu the electric and water were out again. I was able to pack everything into my backpack to reach my goal—only a carry-on bag going home.

A man named Wavimba, who owned an auto and truck parts store in Bukavu, came to take Kathy and me on a driving tour of places I would remember in town. How nice of him to do that. Wavimba spoke pretty good English, so an interpreter was not necessary. We saw the Grace Mission House, on which property my parents and my sister lived for several years, in one of the smaller houses. (This was after they had returned to Africa and left us boys and our families in America.) Then we went to the ruins of the Hotel Riviera, where twenty-eight of us missionaries were held under house arrest by drunk African soldiers with Soviet weapons all the way back in 1960. The hotel was now closed, falling down, and in ruins. Someone was planning to rebuild it, right on the shore of Lake Kivu.

Later, Wavimba took us to see the Orchid Club, which has a restaurant, lakefront lounge, swimming and bar areas. In addition, it is a hotel for outsiders who can afford it. This was an exclusive property built by the Belgians back in the day. Rooms at the Orchid were $300 to $500 U.S. per night, with a few suites that cost even higher. Looking around, we took the steps down to the lakefront. Kathy "sneaked" a cell phone picture of me putting my feet into the lake water. As a kid I had swum in Lake Kivu many, many times. Even though the lake is now giving off gas and carbon monoxide, I stuck my foot in anyway. In fact, there is some concern that the chemicals being emitted may combine to cause a massive explosion in a very populated area. That day things were quiet.

Our tour next took us to the Plaza de la Independence, at the center of Bukavu. As we drove along, a policeman waved us over. No doubt he saw the white people in Wavimba's car, and an opportunity for a bribe. Our driver was hassled over his car's papers. One must have eighteen different permits to maintain a car in Bukavu, Congo DRC. The policeman made him painstakingly drag them all out and show them to him. Finally satisfied, he let us go, after about forty minutes of wondering what would happen next. Amazingly, no money changed hands that time.

On the way to the top of a high hill overlooking the town, we drove past the governor's house—no pictures allowed—and one can only drive by the place. At the top of that hill was a beautiful view of Bukavu, Lake Kivu, the Ruzizi River, Goma, volcanoes, and the horizon all the way into Rwanda. We also saw what I called "Gorilla Mountain." There was a kind of enclosure where some gorillas were housed. With no time to spare, we didn't walk into the gorilla park.

Lastly, we stopped downtown and saw the Grace Mission apartment where my parents and sister lived once upon a time, and the building that was a bookstore and reading room, run by Dad back then. While in the downtown plaza area we were sure we heard a gunshot. Kathy jumped a foot or two, while I was calm and collected—not!

While we were touring, I opened the car door and walked right into a corner of the open door. *Bang*, it hit right on my left cheek. Of course, as an old person I bled profusely and looked like I had lost a boxing match.

Kathy and I also stopped at a local grocery store in downtown Bukavu. I purchased two African shirts to take home with me. We also went to a small religious store to buy some artifacts.

Late afternoon, the Berean pastors and two youth pastors came again with pastor Kitoto. We had promised them that I would give them my thoughts on what I had seen downcountry and at the Kaziba orphanage. They indicated they wanted to start an orphanage in Bukavu, with fifty kids in five locations. True to form, they thought surely I could fund and help with this project. The meeting took about two hours, and then the delegation gave me a three-page letter to take to American churches. Their letter asked for support. After they had left, I realized that I had not made it crystal clear to them that I did not have the resources—neither funding nor human resources—to help with the project. I continued to receive emails from one of the youth pastors for two years after my visit. The ways of the Africans just haven't changed. They wish for the good old days, when missionaries came to take care of them. So sad that due to conditions politically and militarily in Congo DRC, those days are over.

For dinner we shared bread, cheese, Grenadine drink (yum), tomatoes, pineapple, and peanut butter and jelly sandwiches. I sent texts to my kids telling them I'd be headed home the next day, then went to bed.

CHAPTER THIRTY-NINE

A "TOILET TSUNAMI"

THIS WAS THE day I was flying home to Texas. I awoke pretty excited, and with a mixture of nostalgia and sadness that I was leaving Congo DRC. My series of flights home would take the better part of twenty-four hours, and I would "lose a day" on the way home, because of the International Date Line.

Just before the time for me to leave Bukavu, four Balega tribal members came to see me and wish me safe travels. Then Tom and I left his apartment by taxi around 12:30 p.m. local time. We were on our way to the Congo DRC immigration offices at the border crossing. Tom helped me with customs and immigration, including on the Rwanda side. It was so important to have a "protocol" like Tom who knows the ropes and can smooth out the red tape. Rwanda wanted to look through my backpack, so while that took place Tom got a taxi for me to take to the airport at Kamembe, Rwanda. The taxi ride was about $5.50 U.S. (6,000 francs in Rwandan money). Again I lucked out with a driver who knew a little English.

When I got to the "airstrip" at Kamembe, I had to wait until 1:30 p.m., sitting outside on the steps of the shack that passes for a terminal. With the tropical sun beating down, I pulled on a hat, waiting for the building to open. Even though this was a remote airstrip in the middle of nowhere, one could not enter the terminal (shack) until two hours before departure time. When it did open, I went through the security scan and came out convinced that the thing was just for show; I don't think it worked at all. No matter, on I went.

My Rwandair flight arrived and then left thirty minutes late—*This is Africa!* There were twenty-eight passengers on the thirty-seven seat Dash 8, mostly Africans. The flight to Kigali, Rwanda, was uneventful but very rough flying. At Kigali International Airport I had about a three-and-a-half-hour wait for my connecting flight. At the two-hour advance for check-in and security rule, I gained access to my gate for KLM's flight to Entebbe, Uganda, and on to Amsterdam.

While I was waiting in Kigali, three interesting things happened. 1) My newly acquired KLM frequent flyer status got me into the Elite Pearl Lounge for free food, drinks, rest, and free wifi. A man came to announce each flight, so no one would miss their plane. 2) As I used the restroom, something crazy happened. When I pushed the toilet flush button it somehow stuck, and water began to overflow onto the floor. I had to high tail it out of there, before being washed away by a "Toilet Tsunami." Anyway, I got out with a sigh of relief. 3) From Kigali airport I could communicate again with my international phone. I texted my kids, and my youngest granddaughter, Reese, texted me "good night" from America. How sweet! Michelle, my oldest daughter, texted she would meet me at the DFW airport the next day. Things were all set. By the way, this was the first day that my family in the States knew I was safely out of Congo DRC, after many, many days.

As I boarded the KLM flight, I met my seatmate, a really nice middle-aged man from Cologne. He owned a hospital equipment company for caring for premature babies. He had just been to a far corner near the Rwanda and Congo DRC border, installing incubators in a rural hospital. An interesting fellow. Our flight attendant was a veteran with KLM, and we enjoyed a nice chat. She had two kids at home and worked part-time. She could name all her flights and destinations after fifteen years of service. Mostly she flew to Amsterdam, Uganda, Rwanda, Accra, Dubai, Shanghai, and Narita, Japan.

After landing thirty-five minutes later for our stopover and crew change at Entebbe International, I had a one-hour layover, sitting on the plane. The flight to Amsterdam was scheduled to be ten hours. I tried to sleep, but only got two hours at best. KLM fed me at least every three hours. Our "dinner" was served at 11:00 p.m. At long last, we arrived at an overcast Amsterdam Schiphol International Airport. Two travel legs down, two to go . . .

CHAPTER FORTY
AMSTERDAM TO DETROIT

AMSTERDAM FOUND ME tired, but only halfway home. So I kept on truckin'. I boarded an Airbus A-330, one of my favorite long-haul planes, for the Detroit leg. The flight was full, and my seatmate was an African lady—by her dress she was Islamic. She told me she came from Somalia. The flight dragged on and on for about ten hours.

Breakfast was served onboard, for the second time that day for me. Again I tried to sleep, in between listening to the onboard music, Alan Jackson's thirty-four #1 hits. Since KLM and Delta are code-sharing partner airlines, it began to feel like I was back on an American plane. Whereas international flights feed you and feed you, and offer free drinks, American airlines gave us juice, an orange, pretzels, and peanuts. (Welcome home, Rog.)

About ten minutes out from our final approach to Detroit, we hit heavy turbulence—sending one of the flight attendants falling down in the aisle. Thank God we made a smooth landing, and the lady was not injured. I walked about a "million" miles to customs in the belly of the Detroit airport. Then followed another long walk to immigration. The U.S. officials questioned me more than they usually do for returning U.S. citizens. They said it was because I had been gone a month and didn't have anything but a backpack, with no checked luggage. Finally I convinced them that I had been on two mission trips and had given away most of my things to the Africans I had gone to help.

After the hassle at U.S. customs, I headed for my gate and final flight home, to DFW.

CHAPTER FORTY-ONE
HOME AT LAST!

AFTER A COUPLE of ten-hour flights, I was more than ready to be home again. One little problem: I was still in Detroit, awaiting my flight to DFW. I must have looked terrible, because when I checked in for my flight the agent said: "Sir, I'm going to do something really nice for you." To my utter surprise, she handed me a free upgrade to first class. (Did I really look that bad?) Gratefully, I settled into the first-class cabin. All the other people up there in the front were businesspeople, busy on their computers. It surely is nice to be retired from the rat race of constant business demands.

As we flew closer to DFW we hit turbulence again, not uncommon in Texas skies. The landing was smooth, though, and I headed for the baggage claim, eager to see my oldest daughter, Michelle. Now I was really heading home. By the time I got to my house it was early evening, so I unpacked my backpack, drank some juice, ate a quick bite, and went to bed for a *l-o-n-g* rest. I didn't get up until "noon-thirty" the next day, and stood in my shower for an hour, after only spit baths for some twenty-five days.

It was great to be home again.

PART FOUR
UNLIKELY RETURN TO CONGO DRC
IMAGES 2012

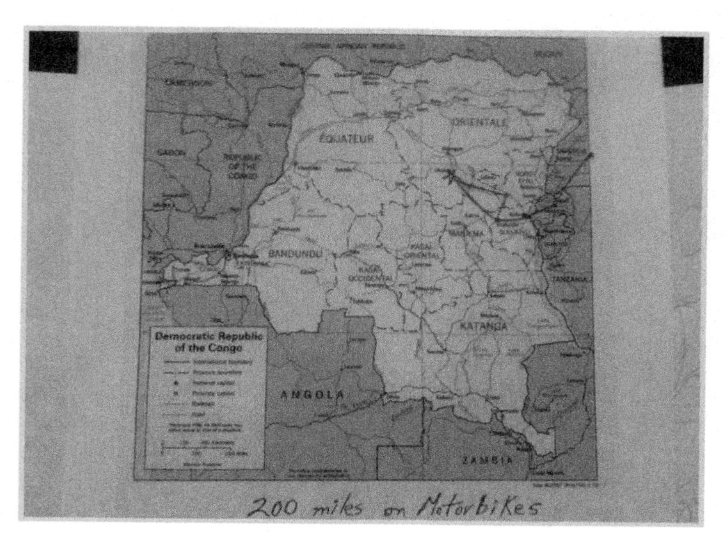

MAP OF DEMOCRATIC REPUBLIC OF CONGO

POLISH "LETT" CARGO PLANE, WE FLEW INTO JUNGLE-2012

Ramazani, my driver for 200 miles in jungle-2012

Tom, Roger, and Kathy ready to travel by "piki" 2012

Rest stop, trail trekking through Congo jungle-2012

Obstacles across the Trail, Congo jungle-2012

Heavy loads in Congo-2012

The jungle ALWAYS wins!

MY HOUSE AT KATANTI, CONGO, BUILT IN 1950S

BACK OF MY HOUSE-2012

Mother's old Singer sewing machine in Congo-2012

Mother's rusting wood stove-2012

TOM LINDQUIST TRANSLATES MY KATANTI GREETINGS-2012

KIBEKIANGABO, OUR FAMILY COOK AGE 93!-2012

My BEST childhood friends, Yoanne and Yosef-2012

Kathy, Tom, SaSimon and Roger: Leaving for USA-2012

PIKI QUIT ALONG TRAIL. THIS IS AFRICA!

ROGER DIPPING IN LAKE KIVU-2012

Elena with Dad's book from 50 years ago-2012

Crowd that welcomed Roger home-2012

RETROSPECTIVE: A DANGEROUS RETURN: SURPRISING LESSONS FROM THE CONGO

"A THIMBLE FULL OF HOPE"

AT THE END of my exciting, unlikely trip to Congo DRC I was amazed at all that had happened to me. Thoughts of the trip still ring in my head to this day, years later. The trip was so totally a "God Thing," supported by many, many people I'd known and loved for years—people like you. Thank you so much for your prayers, love, and support. Not many people get to realize their "lifelong dream" like I did. Thanks be to God. In addition, not many have the opportunity to go and minister in God's name, on behalf of so many back at home.

While the whole thing seemed surreal, it was real. I really did go on mission to Uganda in 2012, and also one time to Congo DRC in 2012. I really got to minister to over seven hundred-fifty kids at the God Cares Summer Camp in Uganda. And I really did get to travel "alone" through Uganda, Rwanda, and into the war-ravaged Democratic Republic of Congo, my home away from home. Traveling by motorbike and rickety cargo planes, I got to revisit my home at mission station Katanti, a place that was originally carved out of the jungle by six dedicated missionaries in the late 1930s! Seriously, my travails and trials in going back "home" were nothing compared to the commitment of the "original six" from Berean Mission.

Having said that, I must add that Congo DRC has gone backward from when I left it in 1960. Every aspect of daily life is much, much harder, more strenuous, and more dangerous. I am saddened by seeing the regression in that country. To me, it has gone back even further than when I first set foot into Belgian Congo in 1950.

On my trip in 2012, I faced dangerous moments and sometimes scary hours traveling and sleeping in the jungles. Sometimes I had no idea what might happen next. But, as I like to say, "God is good *all* the time." I was thrilled to see the places where I grew up as a kid, and more thrilled to see some ninety year olds who still remembered me. And I was privileged to assist in presenting a Biblical Leadership seminar to thirty-five to forty pastors at my "home" church.

The native Balega people in that region have a very difficult life. It is dangerous, and the unexpected is the norm, not the exception. Many, many times the Africans are forced to flee into the forest to escape with their lives, fleeing the tortures, rapings, kidnappings, killings, and burnings of the soldiers who come through the area from time to time. Some had fled into the forest three times just that year, 2012, the year I visited. What an example of commitment. In spite of all of this, these dear people continue to serve the Lord and to further the work begun by the "original six" in 1938, and later by my parents from the 1950s to the 1970s.

Looking at what little they have materially, I was struck over and over by a simple thought. These dear people just need someone to hold out a "thimble full of hope" to them. They take that little bit of hope and turn it into a commitment for the ages. I wish I had such courage.

Many have asked: "So when's your next trip, Roger?" Or, "Would you go back there to live full-time in Congo DRC?" Or, "Are you done with Africa?" I have not felt a call to "full-time service" in Congo DRC. What I have felt is a connection to Congo DRC and Uganda that I will continue to develop. It is such a blessing to continue to return to Africa year after year, serving the millions of orphans in need in Uganda and Congo DRC. There are some 2.5 million orphans in Uganda alone. As God leads me, He will show me how to best minister, serve, and help the people of Central Africa. As I often say: "I cannot change millions of orphans all by myself. However, I can and do affect the life of one orphan at a time." One child at a time . . . that is a privilege beyond measure!

I am a retired salesman, wanting to live out my life "giving back" to the people and places I grew up with in Africa. God has graciously brought my life together with Linda, a dear Christian lady. Linda and I were married in January 2014. Where will this journey take me or us? I'm not so sure, but God surely knows. Stay tuned.

THE END

CPSIA information can be obtained
at www.ICGtesting.com
Printed in the USA
LVOW02s1653200517
534923LV00023B/28/P